ACS SYMPOSIUM SERIES 447

Fourier Transform Infrared Spectroscopy in Colloid and Interface Science

David R. Scheuing, EDITOR
Clorox Technical Center

Developed from a symposium sponsored
by the Division of Colloid and Surface Chemistry
of the American Chemical Society
at the 199th National Meeting
Boston, Massachusetts, April 22–27, 1990

American Chemical Society, Washington, DC 1990

Library of Congress Cataloging-in-Publication Data

Fourier transform infrared spectroscopy in colloid and interface science
  David R. Scheuing, editor

  p.   cm.—(ACS Symposium Series, 0097–6156; 447)

  "Developed from a symposium sponsored by the Division of Colloid
and Surface Chemistry at the 199th national meeting of the American
Chemical Society, Boston, Massachusetts, April 22–27, 1990."

  Includes bibliographical references and index.

  ISBN 0–8412–1895–1

  1. Surface   chemistry—Congresses.   2. Colloids—Congresses.
3. Infrared   spectroscopy—Congresses.   4. Fourier   transform
spectroscopy—Congresses.   I. Scheuing,   David   R.,   1952–   .
II. American   Chemical   Society.   Division   of   Colloid   and   Surface
Chemistry. III. Series

QD506.A1F68   1991
541.3'.3—dc20   90–22913
                                                   CIP

The paper used in this publication meets the minimum requirements of American National
Standard for Information Sciences—Permanence of Paper for Printed Library Materials, ANSI
Z39.48–1984.   ∞

ACS Symposium Series

QD 506
A1 F68
1991
CHEM

M. Joan Comstock, *Series Editor*

1991 ACS Books Advisory Board

Foreword

THE ACS SYMPOSIUM SERIES was founded in 1974 to provide a medium for publishing symposia quickly in book form. The format of the Series parallels that of the continuing ADVANCES IN CHEMISTRY SERIES except that, in order to save time, the papers are not typeset, but are reproduced as they are submitted by the authors in camera-ready form. Papers are reviewed under the supervision of the editors with the assistance of the Advisory Board and are selected to maintain the integrity of the symposia. Both reviews and reports of research are acceptable, because symposia may embrace both types of presentation. However, verbatim reproductions of previously published papers are not accepted.

Contents

Preface

In THE PAST DECADE, IMPROVEMENTS IN infrared spectroscopic instrumentation have contributed to significant advances in the traditional analytical applications of the technique. Progress in the application of Fourier transform infrared spectroscopy to physiochemical studies of colloidal assemblies and interfaces has been more uneven, however. While much Fourier transform infrared spectroscopic work has been generated about the structure of lipid bilayers and vesicles, considerably less is available on the subjects of micelles, liquid crystals, or other structures adopted by synthetic surfactants in water. In the area of interfacial chemistry, much of the infrared spectroscopic work, both on the adsorption of polymers or proteins and on the adsorption of surfactants forming so called "self-assembled" mono- and multilayers, has transpired only in the last five years or so.

Vibrational spectroscopy, through its sensitivity to both intra- and intermolecular interactions, is useful in studies of the "geometric" aspects of molecular packing in both colloidal assemblies and adsorbed layers. Interest remains high—judging from the current literature—in relating the size and shape of surfactant and lipid molecules to those of micelles, bilayers, and vesicles, and in the thermodynamics of colloidal assembly. Increased understanding of the details of intermolecular interactions in such systems will benefit in development of commercial products containing such structures and in comprehension of naturally occurring structures such as cell membranes. The interactions of polymers or proteins with solid surfaces, which may be studied in situ with Fourier transform infrared spectroscopy, affect areas as diverse as lubrication, corrosion, and the development of medical implant devices.

The preparation of a book requires the cooperation of many people. The enthusiastic response I received from both authors and reviewers was especially gratifying, and eased my tasks considerably. I also wish to express my gratitude to the Clorox Company, which provided financial assistance in sponsoring the original symposium. Thanks are also due to all my colleagues at Clorox, especially Jim Rathman and Jeff Weers, for their moral support and encouragement in this project.

Finally, I thank my wife and children for providing the luxury of time and for their support for my work on this volume.

DAVID R. SCHEUING
Clorox Technical Center
Analytical Research and Services Department
Pleasanton, CA 94588

September 14, 1990

Chapter 1

Fourier Transform Infrared Spectroscopy in Colloid and Interface Science

An Overview

David R. Scheuing

Clorox Technical Center, 7200 Johnson Drive, Pleasanton, CA 94588

This chapter reviews the wide range of colloidal systems amenable to investigation by FT - IR spectroscopy. Molecular level information about the interactions of amphiphilic substances in aggregates such as micelles, bilayers, and gels can be obtained and related to the appearance and stability of the various phases exhibited. The interactions of polymers, surfactants and proteins with interfaces, which substantially modify the solid - liquid or liquid - air interface in many important industrial and natural processes, can also be monitored using FT - IR.

Several themes of the application of FT-IR to studies of colloidal particles and interfaces have appeared over the last decade or so. The purpose of this chapter is to attempt to draw together examples of such research topics for both the practicing spectroscopist and workers in the field of colloid and surfactant science. Major advances have been made in recent years in the areas of spectroscopic data handling (which have affected the "look" and "feel" of the analytical laboratory in general) and sample handling (various new optical accessories). These advances have made FT-IR a truly accessible technique, with tremendous potential for application to research in areas of considerable economic, as well as fundamental, importance.

Synthetic surfactants are ubiquitous in the modern world, appearing in a wide variety of consumer cleaning products, in foods and cosmetics, and in industrial products and processes such as metal cleaning and enhanced oil recovery. The design of novel surfactant - based products or processes featuring "improved" performance or decreased environmental impact requires an understanding of the relationship of molecular structure to surfactant phase behavior. In the case of "natural amphiphiles" such as the phospholipids, a single phase, the bilayer, is of paramount importance. Significant advances in the understanding of the molecular forces which stabilize these aggregates have been made recently using vibrational spectroscopy. In nature, membrane bilayers incorporate significant amounts of additional substances such as proteins and cholesterol. Although more complex, such systems are also beginning to yield to spectroscopic analysis. Besides being of fundamental interest, membrane bilayer chemistry affects areas of considerable economic opportunity, considering pharmaceutical chemistry alone.

Many of the spectroscopic concepts developed in studies of bilayers are applicable to other molecular aggregates such as surfactant micelles and gels. The

0097–6156/91/0447–0001$06.25/0

evidence for this statement lies, hopefully, in this chapter, and in the publications of several workers participating in both fields.

Synthetic surfactants and polymers are probably most often used to modify the characteristics of a solid surface, i.e., they function at the solid - liquid interface, such as in the processes of detergency, lubrication, or the formation of adhesive bonds. The performance of modern FT - IR spectrometers is such that many new applications to the characterization of the solid - liquid interface, particularly in kinetics studies, are possible. Reflection - absorption spectroscopy and attenuated total reflectance (ATR) techniques have been applied to "wet" interfaces, even the air - water interface, and have figured prominently in recent studies of "self - assembled" mono - and multilayers.

Interestingly, protein adsorption is also a field of biological interfacial chemistry which parallels that of synthetic materials at the solid - liquid interface. A number of spectroscopic advances have been made which allow FT-IR to be used in kinetic monitoring of protein adsorption on metals and "biocompatible" polymers. In addition to providing in - situ measurements of total adsorbed protein, FT - IR can also yield information about perturbation of protein secondary structure in adsorbed layers.

In the various sections of this chapter, I will briefly describe the major characteristics of FT-IR, and then relate the importance of these characteristics to physiochemical studies of colloids and interfaces. This book is divided into two major areas: studies of "bulk" colloidal aggregates such as micelles, surfactant gels and bilayers; and studies of interfacial phenomena such as surfactant and polymer adsorption at the solid-liquid interface. This review will follow the same organization. A separate overview chapter addresses the details of the study of interfaces via the attenuated total reflection (ATR) and grazing angle reflection techniques.

FT-IR Brief Description

Interferometry. A detailed description of spectrometer designs, of which there are several now commercially available, is beyond the scope of this book. There are several fine texts to which workers new to the field, or interested non - practitioners, may refer.(1 - 3).

Several major points should be mentioned. In FT-IR *interferograms* are recorded, and the infrared spectra *computed* from the interferograms, via a fast Fourier transform algorithm introduced relatively recently (4). It is the replacement of the monochromator of earlier spectrometers by an interferometer which is primarily responsible for the improved performance of FT instruments.

The use of an interferometer, the concept of which was first described by Michelson in 1891 (5), allows all of the radiation interacting with the sample to fall on the detector during all of the time of measurement, i.e., data from all spectral frequencies of interest is collected simultaneously, which yields the so-called multiplexing, or Felgett's advantage (6). The measurement of a single interferogram can be made rapidly, resulting in a large savings in time of measurement. As is typically done in chemical spectroscopy, interferograms are repeatedly obtained and averaged, to yield spectra of very high signal to noise ratio (increasing with the square root of measurement time, or number of scans, assuming the typical case of constant velocity of the moving mirror of the interferometer).

Jacquinot's advantage refers to the improved signal to noise ratio in spectra measured with interferometry due to the greater signal size present at the detector. Griffiths (2) points out that the magnitude of Jacquinot's advantage in modern spectrometers had been exaggerated in earlier times. Theoretical calculations which indicate 200 to 2000 - fold sensitivity advantages for FT- IR are somewhat misleading, since overall spectrometer performance, in the case of FT instruments, must take into account limitations imposed by detector design, detector foreoptics, and beamsplitter efficiencies. In many cases, instrument performance is limited not by optical throughput, but by the dynamic range of the analog to digital converter used, since there can be an enormous difference in signal intensity between the centerburst and in the wings of the digitized interferograms being recorded (1).

Given that the multiplexing advantage can be used to obtain spectra of samples exhibiting large energy losses, such as aqueous solutions, thin films, or powders, another aspect of FT-IR also emerges as important. The precision of the digitization intervals of the interferograms, and hence in the computed spectra, is extremely high.   Laser light fringing methods are commonly employed in monitoring the position of the moving mirror in the Michelson interferometers used in modern FT-IR spectrometers. The precision with which resulting infrared bands are digitized is such that it can be ignored as a source of error in the determination of the frequency of many bands in the spectra of condensed phase samples. The determination of the frequency of a Lorenztian band is limited, for practical purposes, by the signal to noise ratio of the spectrum (7). In the case of a signal to noise ratio of 1000, which is often readily achievable, the precision of band frequency determination is < 0.01 cm^{-1}, for a band recorded at 4 cm^{-1} nominal resolution. Such results certainly suggest, taking such band frequency (and width) information into account, together with interpretations of band shifts, that FT-IR can indeed be an "information - rich" technique.

Spectral Manipulation Techniques.   Many sophisticated software packages are now available for the manipulation of digitized spectra with both dedicated spectrometer minicomputers, as well as larger main - frame machines. Application of various mathematical techniques to FT-IR spectra is usually driven by the large widths of many bands of interest. Fourier self - deconvolution of bands, sometimes referred to as "resolution enhancement", has been found to be a valuable aid in the determination of peak location, at the expense of exact peak shape, in FT-IR spectra. This technique involves the application of a suitable apodization weighting function to the cosine Fourier transform of an absorption spectrum, and then recomputing the "deconvolved" spectrum, in which the widths of the individual bands are now narrowed to an extent which depends on the nature of the apodization function applied.   Such manipulation does not truly change the "resolution" of the spectrum, which is a consequence of instrumental parameters, but can provide improved visual presentations of the spectra for study.

The "price" paid for deconvolution of spectra is an increase in the noise level, as well as the potential for producing "ringing" baselines or subsidiary lobes on the wings of real spectral features. In the development of the Fourier deconvolution technique over the past several years, considerable efforts have been made in investigation of the effects of signal to noise ratio, selection of the weighting function used for deconvolution, and comparisons of this approach with that of even - order derivative manipulations (8-13). A recent demonstration that peak areas are practically unaffected by proper application of deconvolution techniques suggests that quantitative measurements employing deconvolved spectra are possible (14).

Other manipulations of spectra are also possible, and are used with varying degrees of success. Discussions of curve fitting (15), factor analysis (16,17), derivative formation and smoothing (18) have all appeared. Curve fitting has been applied to several weak bands in the spectrum of a micellar surfactant solution (19), and work on improvements in fitting second derivative spectra has continued (20). As stated in an article on the origin of artifacts in deconvoluted spectra, however, one must avoid the pitfalls in mathematical manipulation of spectra   "... in order to remain in the field of spectroscopy without entering that of spectrology." (21).

Difference Spectroscopy. Spectral subtraction, or difference spectroscopy, is the final important topic of this general description. In studies of "bulk" samples such as micelles, bilayers, or vesicles, water (or D_2O), which is an intense absorber throughout most of the mid-infrared spectrum (4000-400 cm^{-1}), is the solvent. The subtraction of the spectrum of liquid water from that of an aqueous solution of an analyte of interest can be accomplished routinely with many of the dedicated minicomputers used with modern FT-IR spectrometers. The spectrum of liquid water is recorded under the same conditions (cell pathlength, temperature, cell window type) as the samples of interest, and digitally stored. This reference spectrum of water can then be scaled by an appropriate multiplicative factor (usually near 1.0) and subtracted from the sample spectrum, to produce a spectrum of the analyte of interest (22,23). Residual noise in the difference spectra of aqueous samples of high water content often persists above 3100 cm^{-1}, in the region of the intense H-O-H stretching bands, which is usually not too severe a limitation. Successful subtraction of the H-O-H bending band of water near 1640 cm^{-1} is usually of more concern, because this broad band overlies the region where important bands due to ester, amide, and carboxylate groups occur. Limiting the pathlength of the cell employed to less than 25 micrometers aids in controlling the intensity of this band, and improves subtraction results in this frequency range. The necessity for working with such relatively short pathlengths, which are accomplished in ordinary transmission cells through the use of thin gaskets, prompted the development of several optical accessories base on attenuated total reflection (ATR) optics. Infrared radiation propagating inside a multiple internal reflection element (IRE) of an ATR accessory interacts with the external medium (sample solution) to a distance of only 1 micrometer or so. By adjusting the dimensions of the IRE, and hence the number of internal reflections, an ATR accessory can provide an effective pathlength from 10 to 20 micrometers, a range which is ideal for use with aqueous solutions. Several designs, in which the IRE is conveniently located within an easily filled chamber, have appeared (24,25). Studies of the adsorption of proteins from aqueous solutions, which necessitated the subtraction of the water bending band, have been conducted using ATR sampling optics (26). The adsorption of proteins onto an IRE surface also points to a potential pitfall in the use of such accessories in studies concerned with "bulk" samples. Adsorption of a surfactant onto the IRE will provide spectra containing information about both bulk and adsorbed species, and this potential distortion of the spectra should be avoided in physiochemical studies of bulk phases.

Subtraction of the spectrum of liquid water, even of moderate band intensity, can also be complicated by solute-water interactions which cause a shift in the H-O-H bending bands, making a complete nulling of the band in the difference spectrum impossible (23). As discussed further below, in bulk phase samples such as microemulsions or inverse micelles of moderate water content, significant information about aggregate structure is obtained from shifts in the water bands.

Spectral subtraction can also be used to enhance the more subtle differences between two samples by the nulling of spectral features common to two spectra, which leaves the differences between samples as excursions from an otherwise featureless baseline. Interpretation of such difference spectra is usually done in conjunction with measurement of other spectroscopic changes, such as band shifts or intensity changes, that are produced in a series of spectra of a bulk phase sample perturbed in some manner. Examples will be discussed further below in the various subsections.

Characterization of Colloidal Aggregates

Bilayers. A considerable body of literature exists on the application of FT-IR (and Raman) spectroscopy to characterization of the phase behavior of both naturally occurring and synthetic phospholipids. Such compounds have the structure typified by   dipalmitoylphosphatidylcholine

(DPPC);

These materials are of keen interest in biochemistry because of their aggregation into bilayers or multilamellar vesicles, which serve as models of cell membranes. Of course, these compounds can also be viewed as double - tailed zwitterionic surfactants, with a tendency to form aggregates such as bilayers that resembles the behavior of many commercial synthetic surfactants (single- or double-tailed) in certain areas of their phase diagrams. The interplay between the packing of the hydrophilic headgroups and the hydrophobic tails of lipids is a major factor determining the phase behavior of lipids. A molecular packing model developed by Israelachvili et.al. (27) provides a simple, but effective geometrical model of aggregate shape which explicitly treats the balance between heagroup area and cross - sectional area of the hydrophobic tails. An examination of the application of FT-IR to lipid bilayer characterization can thus be illuminating to workers concerned with the phase behavior of synthetic surfactants or mixtures of surfactants. Recent work suggests that some mixtures of synthetic surfactants spontaneously adopt vesicular arrangements (28).

Methylene Chain "Tail" Packing. Amphiphilic compounds contain the common structural feature of a hydrocarbon tail, which provides a key spectroscopic link to alkanes, which have been studied in great detail (29-40). The frequency, width, peak height, and integrated intensity of the intense asymmetric and symmetric CH_2 stretching bands, near 2925 and 2850 cm^{-1}, respectively, are sensitive to the gauche/trans conformer ratio of the hydrocarbon chains. Thus, these vibrational modes can be used to monitor changes in the ratio as the "ordering" of the tails of amphiphilic compounds changes, in response to temperature, pressure or composition of the aggregate.

DPPC bilayers, as an example, undergo a major transition at a temperature of 41.5 °C. This transition involves a large increase in the number gauche conformers in the tails of the lipid, corresponding closely to the melting of a hydrocarbon, which is known as the gel to liquid crystal transition, often denoted as T_m (see reference 41 and references therein). Over the narrow temperature range of 40 to 42 °C, the symmetric CH_2 stretching band of DPPC is thus found to shift from near 2850 cm^{-1} to near 2852 cm^{-1}. An increase in bandwidth is noted as the melting of the chains at the gel to liquid crystal transition introduces gauche conformers. This change in the packing of the lipid tails of a bilayer structure is a biologically important response of bilayer interior fluidity to temperature. The phase behavior of the lipid tails of bilayers is actually even more complex than this. As the temperature is reduced below T_m, changes in the width of the CH_2 deformation band ("scissoring" mode) near 1470 cm^{-1} are noted (41). The width of the band increases as the temperature is lowered, eventually resulting in the production of a doublet. Such "factor-group splitting" of the CH_2 scissoring and rocking bands is due to interchain interactions when the chains are packed in an orthorhombic subcell, as work on the phase behavior of solid hydrocarbons has shown (36). Temperature - dependent distortion of the orthorhombic subcell adopted by the lipid tails can be monitored by the changes in the scissoring band, and a "pretransition" temperature defined. The thermotropic phase behavior of the lipid tails is found to resemble the surprisingly complex phase behavior of longer chain solid alkanes of odd carbon number. These alkanes exhibit a solid state transition from a relatively "ordered" orthorhombic phase to a more "disordered" hexagonal (rotator) phase as the melting point is approached (29). The temperature - induced changes in molecular packing in aggregates such as lipid bilayers are intimately connected with chain defect formation. Increasing the gauche conformer content of the tails increases their cross sectional area. In accordance with the geometrical constraints on amphiphile packing for a given aggregate shape (27), appropriate rearrangements in the packing of the headgroups might be expected to occur for maintainence of the bilayer configuration. As discussed below, headgroup bands of amphiphiles can serve as monitors of molecular packing in aggregates as well. The ability of FT-IR to provide molecular level information simultaneously about both amphiphile tails and heads is of considerable importance.

Headgroup Region Structure. The C=O band of the lipid ester groups is an example of a useful headgroup band. Shifts in this broad band, which occurs near 1734 cm^{-1}, are observed at both T_m and at the pretransition. Extensive dehydration of the ester groups occurs at lower temperatures (42), defining a "subtransition". Fourier deconvolution of the ester band reveals that it is comprised of two bands, assignable to the sn - 1 and sn - 2 ester moieties, which change in relative intensity, but do not shift in frequency, as the temperature is changed. The ester band of sulfocholine lipids was employed, together with the CH$_2$ stretching and deformation bands, in a comparison of the subtransition kinetics of DPPC and DPPsulfocholine (43). Addition of cholesterol to DPPC bilayers was found to alter packing of the lipid in the headgroup region, inducing inequivalance between the ester carbonyl groups, and altering T_m.

Two other lipid headgroup bands found in the spectra of fully hydrated lipid bilayers are the O=P=O asymmetric and symmetric stretching bands of the phosphate group near 1120 and 1080 cm^{-1}, and the (CH$_3$)$_3$N stretching band near 970 cm^{-1} (41). The effects of hydration and sampling handling on all bands of phosphatidylcholine, phosphatidylethanolamine, and mixtures thereof have also been discussed (45). The broad phosphate band is usually unaffected by changes in temperature (44,46), reflecting a constant extent of hydration. However, in a study of phosphatidylserine (PS) vesicles, the bidentate complexation of Ca^{+2} ions to PS results in a dramatic splitting of the symmetric O=P=O stretching band into four components, due to the "local site" symmetry lowering of the phosphate group complexed with Ca^{+2} (47). The effect of Ca^{+2} on the phase behavior of PS is extensive, as detected by changes in the ester C=O band profile and the elimination of T_m. Such dramatic effects of headgroup environment on the packing of lipid tails in bilayers are consistent with the results of other studies, in which the effect of headgroup type on tail packing of various phospholipids were examined (48,49). The temperature - dependent splitting of the CH$_2$ scissor band of lipids with bulky choline headgroups was always distinct from that of other lipids with smaller amine or acid headgroups. Smaller headgroups allowed tighter packing of the tails, resulting in bands which more closely resembled those of alkanes, in which the perturbing headgroup is absent. These studies also indicated that perturbation of the tail packing increases as the tail length is reduced, consistent with interpretations of data from other techniques. Inter - chain interactions are apparently largest near the chain centers, and decrease toward chain ends.

The frequency of the (CH$_3$)$_3$N band has been found to be sensitive to the extent of hydration of the headgroups, exhibiting shifts upon the formation of ice (41), and with incorporation of cholesterol into the bilayer (44).

Quantitative Structural Descriptions of Bilayer Structure. The frequency and shape changes exhibited by the CH_2 stretching bands at T_m can be employed as a quantitative tool in describing the cooperative chain melting process in binary phospholipid mixtures (50). A mixture of DPPC and perdueterated dimyrystoylphosphatidylcholine (DMPC-d_{54}) allows independent monitoring of the extent of melting of the chains of both lipids, because of the separation of the CH_2 (2852 cm^{-1}) and the CD_2 stretching bands (2098 cm^{-1}). A two-state model can be applied to the gel to liquid crystal transition, in which the CH_2 (CD_2) stretching frequencies in the spectra of the gel and liquid crystal phases represent the thermodynamically sensitive parameters. In the two - phase region across the temperature - induced transition, the spectra obtained are assumed to be linear combinations of the spectra of the gel and liquid crystal phases. The extensive overlap of the CH_2 (CD_2) bands in the spectra of the two phases introduces a complication, however, which was first noted in FT-IR studies of the micellization of alkanoates (51). The change in frequency of the CH_2 band is a smooth, but non-linear function of the extent of transition, and calculations of the extent of transition must take this into account. A combination of the spectroscopic data and calorimetric measurements of the enthalpy changes allowed application of Zimm - Bragg theory, and supported the conclusion that compositional segregation of the lipids in the bilayer was present.

One of the major goals of these many investigations of lipids is, of course, a better understanding of the in - vivo behavior of membranes. Beyond studies of binary lipid mixtures, as mentioned above, a further step which is necessary is the incorporation of proteins into the layers. In many respects, this increase in the complexity of the bilayer systems resembles that encountered in the use of synthetic surfactants in "real - world" situations, where blends, rather than single, surfactants are used. Surfactant blends in aqueous solutions are often further modified in use by the solubilization of oily organic compounds, as in the cases of detergency or cosmetic formulation.

FT - IR has been used successfully to develop the phase diagram of binary lipid bilayers with incorporated proteins, which agreed with those developed by differential scanning calorimetry (52). Depending on the type of lipid and protein, selective or non - selective perturbations of T_m can occur. In addition, the miscibility of lipids in the gel phase can be investigated, and related to the type of protein present (53,55). Incorporation of protein into bilayers is thought to affect the protein secondary structure. Deconvolution of the broad amide I carbonyl band between 1700 and 1600 cm^{-1} into several bands known to be sensitive to the relative proportions of α helices, β sheets, and turns is possible. Substantial changes in the protein secondary structure upon binding to lipids is detected (56).

Bilayers of Synthetic Materials. The development of the FT-IR methods outlined above has led to the application of the technique to bilayers of synthetic materials which are of interest in industrial applications, such as drug delivery systems. The interaction of gramicidin A, a polypeptide, with bilayers of dioctadecyldimethylammonium bromide and DPPC causes changes in T_m of the bilayers, as monitored by the CH_2 bands of the amphiphile. Differences in the site of incorporation of the polypeptide in the two types of bilayers (57) can be detected. The differences in hydrogen bonding capability between dimyristoylphosphatidylcholine (DMPC) and an analogue where an amido linkage replaces the ester linkage is responsible for significant differences in thermotropic behavior. The CH_2 stretching bands of the amido compound indicate gradual tail disordering over an 18 °C range in temperature. Difference spectra in the C-H stretching region and changes in the amide II (1558 cm^{-1}), antisymmetric P=O stretching (1214 cm^{-1}), and the C-N stretching (970 cm^{-1}) are used to establish the presence of three transitions of the amido compound between 10 and 40 °C (58). The FT-IR spectra of another unusual lipid, 1,2-bis (10,12 -tricosadiynoyl)-*sn-*

glycero-3-phosphocholine, indicated that both helical and tubular structures formed by the compound are composed of similar lipid microstructures (59). The persistence of strong intermolecular hydrogen bonding in another amide functional double-tailed amphiphile across the gel to liquid crystal transition was demonstrated by an FT-IR study, employing shifts in the CH_2 stretching, C=O stretching, and amide I band (60).

Lamellar Phases of Surfactants

The sensitivity of the CH_2 stretching and deformation bands to the extent of disorder in the hydrocarbon tails allows studies of the coagel to gel and gel to liquid crystal transitions of surfactants. A comparison of the thermotropic behavior of octadecyltrimethylammonium chloride (ODAC) and the dioctadecyldimethyl compound (DODAC) at high surfactant concentrations (18 - 21 % water) was made (61,62). As the temperature is increased the CH_2 stretching bands of both surfactants exhibit two sharp frequency increases. The coagel to gel transition is followed at higher temperature by the gel to liquid crystal transition. The thermotropic behavior of the CH_3 stretching bands of the CH_3-N^+ headgroup, however, suggests that in the gel phase of DODAC, the hydrated headgroups are much more highly ordered (as is a significant fraction of the water) than in the case of the single tailed ODAC. Employing polarized IR radiation, further studies of water - DODAC coagel and gel phases elucidated the headgroup packing arrangement in these phases, and confirmed the highly ordered headgroup packing in the gel phase (63).

Micelles

Micellar solutions of amphiphilic compounds, both normal (water continuous) and inverse (oil continuous), continue to be a major topic of colloid research. Use level concentrations of surfactants are often low enough that only the micellar phase is present. Therefore, the relationship of micelle structure (aggregation number and shape, charge, etc.) to aspects of performance (solubilization capacity and rate, adsorption processes) remains of keen interest. Fundamental studies of aqueous micellar solutions continues, in an effort to better understand the liquid-like hydrocarbon interior, and the aqueous environment of the headgroup region (for example, see ref. 64). Micellar shape changes, i.e., the sphere to rod transition, are also of interest, because of the unusual rheological properties of rod- or thread-like micelles (for example, see ref. 65). FT-IR has been applied to characterization of the processes of micelle formation and a wide range of micelle properties.

Normal Micelles CMC and CMT Transitions. The concentration - induced monomer to micelle transition of C5 to C10 sodium alkanoates (51) was monitored using the frequencies and widths of the asymmetric CH_3 and CH_2 stretching bands, and the antisymmetric COO- stretching bands, yielding information about both tail and headgroup environment changes. A rather sharp decrease in the frequency of the CH_2 band is noted as the concentration of the C8 and C10 alkanoates is increased above that typically accepted as the critical micellization concentration (cmc). The transfer of the hydrocarbon tails from an aqueous environment in the monomeric state to the micelle interior is responsible for the frequency shift. As the tail length is deceased from C8 to C6, the concentration range of the shift becomes considerably broader, as the cmc becomes more ill-defined. The carboxylate band (near 1550 cm^{-1}) exhibits a shift toward higher frequency at the cmc as well, due to the necessary increases in counterion binding accompanying micelle formation (23).

These early studies also showed that the sharp change in the CH_2 band frequency at the cmc can be explained by a two-state model. The linear combination of two synthetic Lorenztian bands, separated in frequency by much less than their total widths, was shown to give rise to non-linear plots of frequency *uersus* band intensity ratio, in qualitative agreement with the frequency - concentration plots of C8 and C10 alkanoates. In the case of C5 and C6 alkanoates, the frequency shifts could not be modeled with a two state system, leading to the interpretation that these molecules undergo pre-micellar aggregation.

Another important transition of surfactants involving micelles, the critical micellization temperature (CMT), has been found to be readily amenable to study by FT-IR, largely because of the relatively high surfactant concentrations involved (>0.1 M). The CMT is concentration dependent up to concentrations of about 0.1 to 0.3 M, above which the dependence decreases significantly. The Krafft point is thus found at lower temperatures than the CMT, and can be considered the CMT at the cmc (63-65). A thermostatted transmission cell for control of the temperature of the surfactant solutions, held between CaF_2 or BaF_2 windows, is necessary. Automation of the entire spectroscopic CMT experiment has been described (66).

The change at the CMT involves conversion of a poorly hydrated solid surfactant coagel to an isotropic micellar solution. The spectra thus exhibit large frequency increases for the CH_2 bands accompanying the "melting" of the surfactant tails over a narrow temperature range, suggestive of a highly cooperative process (63). The CMT is a strongly increasing function of hydrocarbon tail length, and exhibits a sensitivity to unsaturation in the tail (65). Since changes in headgroup hydration are involved in the transition, the CMT also is sensitive to the nature of the headgroup (for example, alkyl sulfate CMTs < carboxylate CMTs), and the nature of the counterion (63). The headgroup bands due to asymmetric and symmetric SO_3 stretching in sodium hexadecylsulfate exhibit significant changes at the CMT, due to increased hydration and changes in the location of the mobile sodium counterions. In the case of alkanoates, the asymmetric and symmetric COO^- stretching bands respond differently at the CMT, yielding information about sodium counterion location (63).

Normal Micelles - Solubilizate Probes. The addition of a probe molecule, usually bearing a C=O group, to a micelle has been used to asses the solubilization site of the probe (67) and to infer the extent of penetration of water into micelles (68,69). The basis of such studies is the well known decrease in the C=O band frequency upon hydrogen bond formation (70 -73). Two important concepts must be addressed, however, when using probes in studies of micelles: the solubilization site of the probe (micelle core or palisade layer); and the possibility of probe-induced changes in the micelle.

The use of a "ketone-labelled" surfactant as a probe (7 - oxooctanoate) of water penetration into sodium octanoate micelles yielded spectra exhibiting a single ketone C=O band at 1689 cm^{-1}, characteristic of hydrogen-bonded C=O groups, which suggested penetration of water molecules up to the seventh carbon atom, i.e. deep into the micellar interior. Using 5 - nonanone as the probe, three carbonyl bands are found, with one at 1720 cm^{-1} indicating the presence of non-hydrogen-bonded C=O groups. Thus, the lack of a carboxylate headgroup allows this probe to reside in two environments, which apparently differ in water content. These types of studies illustrate an advantage of FT-IR over NMR spectroscopy. If two species exhibiting distinct infrared spectra are present in detectable amounts, two bands will be observed, without interference from chemical exchange, which in the NMR experiment results in an "averaging" of the environments detected by the probe.

A full discussion of water penetration into micelles is beyond the scope of this chapter. The results described above, and others employing longer chain keto - surfactants in other micelles and bilayers, indicated a trend toward less water penetration to the core of aggregates as the surfactant tail length increased, and as aggregate curvature decreased (bilayer formation). More data from FT-IR studies employing probes, perhaps combined with information from an ancillary technique such as light scattering obtained on the same micelle, will be necessary to fully incorporate FT-IR results into a model of micelle structure and micelle dynamics.

Effect of Pressure on Micelles. While temperature studies of the phase transitions of bilayers and micelles have been performed for some time now, the utilization of pressure as a variable is a more recent development. Variation in temperature of a colloidal aggregate such as a bilayer causes simultaneous changes in thermal energy and volume, whereas isothermal variation in pressure (up to 50 kbar) yields spectroscopic changes due only to volume effects. A review of high pressure vibrational spectroscopy of phospholipid bilayers has recently appeared (74), in which the surprisingly rich barotropic phase behavior of these compounds is explored in detail.

In a study of pressure effects on sodium and potassium decanoate micelles, the frequency of the CH_2 stretching band was found to increase with pressure, with a discontinuous drop at the critical coagelization pressure (75). These results indicate that external pressure applied to micelles induces hydrocarbon tail disordering, even at pressures as high as 20 kbar, followed by a large increase in tail ordering upon coagel formation.

The critical coagelization pressure is affected strongly by the nature of the counterion, shifting up by 1.3 kbar when sodium is replaced by potassium. The carboxylate stretching bands are sensitive to counterion type and packing in the coagel phases. The smaller sodium counterion is bound tightly to the carboxylate headgroups of the soap. Weakening of the symmetric COO– stretching band intensity, due to the reduction in the transition moment of this mode, is observed. The tighter binding of sodium allows greater reorientational fluctuations of carboxylate groups, even in the coagel phase, with a resultant difference in aggregate structure between the sodium and potassium soaps.

Pressure experiments with sodium octanoate micelles containing a probe molecule (sodium 7 - oxooctanoate or 5-nonanone) have been described (76). Changes in the extent of hydrogen bonding of the C=O groups of the probe as a function of pressure were observed, and attributed to removal of water from the micellar core at the coagelization pressure. A change in the solubilization site of the ketone from the micelle interior to the palisade region at pressures well below the transition was noted.

Mixed Micelles. A considerable amount of interest in micelles containing two (or more) dissimilar surfactants exists (77) because in many practical uses of "formulated" products, mixtures are usually present. Synergistic increases in viscosity and solubilization are sometimes observed in mixed micelles, and are attributed to the non-spherical shapes frequently adopted by such mixed micelles. The application of FT-IR to the characterization of mixed micelles has not been extensive, but shows promise in elucidating mixed micelle structure (78). Evidence for increased crowding of the hydrocarbon tails of a cationic (dodecyltrimethylammonium chloride) and an anionic (sodium dodecylsulfate, SDS) surfactant in mixed micelles was found, based on the frequencies of the CH_2 bands as a function of micelle composition, which were correlated with changes in the aggregation numbers of the mixtures measured by an independent technique. Simultaneous changes in SO_3 and CH_3-N^+ headgroup bands were also found, and could be attributed to the strong electrostatic interactions of the surfactants, at the expense of interactions with sodium counterions.

Mixed micelles of an amine oxide and SDS exhibit extremely diverse rheological behavior as a function of composition within the water-rich region of the phase diagram of the system commonly referred to as the L_1 micellar phase. The spectra show a close correlation of increased tail (and headgroup) ordering and increases in solution viscosity. A comparison of the spectra of the hexagonal phase formed by the mixture at higher total surfactant concentration with those of the most viscous "L_1" samples showed remarkable similarity. The interactions between the surfactants in the viscous L_1 samples and in the hexagonal phase were interpreted as being very similar. The long, thread - like mixed micelles of the viscous L_1 phase were responsible for the rheology observed (79).

Inverse Micelles

The structure and dynamics of inverse (water in oil) micellar solutions and microemulsions are of interest because of the unique properties of the water core, the view that such micelles may serve as models of enzyme active sites, and the potential use of inverse micelles as hosts for enzymatic reactions (80-82).

The monomer to inverse micelle transition of dodecylammonium propionate (DAP) in CH_2Cl_2 was monitored by the shifts in the COO^- bands. Addition of water caused further decreases in the COO– band frequency due to headgroup hydration. Micelle formation in hexane yielded similar spectroscopic changes (83). Frequency shifts induced by hydrogen bonding to water in the inverse micelle core were illustrated for phosphonate, carboxylate, and sulfate headgroups (84).

Characterization of the water in the inverse micelle (or microemulsion) core is of interest, and attempts at distinquishing "bound" (to the headgroups of the surfactant) and "bulk" water through shifts of the O-H stretching and bending bands have been made. In an early study using a grating spectrometer, reverse micelles of sodium octanoate/decanol/water were studied (85). The average energy of hydrogen bonds formed by water was monitored by the frequency of the asymmetric O-H stretching band of water near 3500 cm^{-1}, which decreases in frequency as the hydrogen bond strength increases. This study showed that hydrogen bonding of water in decanol alone increased as the water concentration increased. However, in the ternary system, the hydrogen bond strength decreased as water was added to the system, at constant octanoate/water ratios. The importance of the presence of both hydrated sodium ions and carboxylate groups to the water structure was thus illustrated.

Inverse systems formed by Aerosol OT (AOT,bis(2-ethylhexyl) sodium sulfosuccinate) are probably the most well characterized. In a study aimed primarily at characterizing the structure of water in water/AOT systems containing heptane, CCl_4 and toluene as the oil phase, a curve - fitting technique was used to fit the O-H stretching bands of water (86). The O-H stretching band of water in the spectra of the inverse systems could be fitted by three Gaussian bands, located near 3580, 3480, and 3280 cm^{-1}, while those of liquid water could be fit by the sum of two major bands at 3455 and 3280 and a very minor band near 3600 cm^{-1}. A subtraction of the spectrum of liquid water from that of the water in the inverse system produced an O-H stretching band which could be fit with two components at 3580 and 3480 cm^{-1}. These bands are interpreted as arising from "interfacial" water molecules in the core. Increasing the molecular ratio of water to AOT in the systems increases the size of the micellar core. The total area of the O-H bands increased linearly with water content, but with a change in slope at a water / AOT ratio above 4.4. The changes in the ester C=O bands and the symmetric S-O stretching bands, as functions of water content, respond similarly. All these changes suggest that the inverse micelles swell to form a microemulsion when more than 4 water molecules per surfactant headgroup are present. The addition of salt (1 - 8%) to the water core was found to have no effect on the water bands, which confirm findings by other techniques that salt addition decreases intermicellar attractive forces without changing micelle structure or size.

Since the structure and size of AOT inverse micelles can be controlled, a recent demonstration of the use of FT-IR in monitoring the lipase - catalyzed hydrolysis of triglycerides in inverse systems has appeared (87). Inverse micelles of AOT at 50 mM in isooctane, with a water/surfactant molar ratio of 11.1 were prepared. The presence of triglyceride in the isooctane was found not to affect the micelle size, and the C=O bands of the triglycerides were unaffected by the presence of AOT. Partitioning of fatty acids between the isooctane and the micelles was readily detected by shifts in the O-H stretching band of the acid groups from 3041 cm^{-1} in isooctane to near 3500 cm^{-1} in the presence of AOT micelles, and by the decrease in intensity of the acid C=O band near 1715 cm^{-1}, as a function of the number of micelles present. The hydrolysis of a triglyceride was monitored by the decrease in the intensity of the water O-H band near 3457 cm^{-1}, a decrease in the ester C=O (1751 cm^{-1}) or C-O (1161 cm^{-1}) band intensities, or the increase in the C=O band of the fatty acid products (1715 cm^{-1}). The enzymatic reactions of a variety of lipases with trioctanoin, and the dependence of the hydrolysis kinetics on the water content of the inverse micelles were investigated.

Other types of inverse micelles, such as those of egg yolk phosphatidylcholine (87) and a calcium alkylarylsulfonate (88) have also been investigated. As recently discussed (88), the properties of inverse micelles formed by different surfactant molecules should be expected to be quite diverse. As an example, the interaction of a solubilizate (TbCl$_3$) with surfactant headgroups of the calcium sulfonate surfactant mentioned above was found by FT-IR and energy transfer methods to be dramatically slower (10-150 hr) than that found with AOT (milliseconds).

The use of inverse micelles and microemulsions of AOT in supercritical or near supercritical fluids as extractants for valuable hydrophilic substances such as proteins continues to develop. FT - IR studies of the pressure dependence of the water core structure in various parts of the phase diagrams of such systems have been described (89).

The aggregation of the series of dodecyl ethoxylates (C12EO$_n$, n=1,2,4,5,8) in aliphatic hydrocarbons, benzene, and CCl$_4$ was determined from an infrared study of the O-H stretching bands (90). Monomer concentrations were calculated from the intensity of the O-H band near 3500 cm^{-1}, which has been assigned to intramolecularly hydrogen-bonded hydroxyl groups. Quantitative determinations of the aggregation of C12EO$_8$ in heptane yielded aggregation numbers in agreement with those determined by independent methods.

Microemulsions

A major goal in preparation of this volume is the elucidation of the potential of FT - IR to studies of a wide variety of colloidal aggregate structures. The partially unrealized potential of the technique is perhaps responsible for the lack of application to the characterization of the wide variety of systems known as oil in water microemulsions. The structure of such systems is a topic of keen interest, since depending on the surfactant and oil identities, either a droplet or a bicontinuous structure (or variety of structures) can account for experimental observations. The compositions of many of these systems are such that reasonable concentrations of surfactant, oil, and sometimes a co- surfactant such as an alcohol are present, suggesting that characterization of some or all of the components by FT - IR should not be limited by sensitivity issues. On the other hand, band overlap would be expected to be a problem in such spectra, which would need to be addressed by the demonstrated proper application of difference spectroscopy, and possibly with the use of selectively deuterated components. In unpublished studies in our laboratory, for example, we have found that the extent of hydrogen bonding of the ethylene oxide groups of alkyl ethoxylate surfactants in ternary systems of hydrocarbon/surfactant/water can be assessed by shifts in the intense C-O stretching bands of the EO groups.

The application of NMR spectroscopy to the determination of order parameters of surfactant tails in microemulsions and liquid crystals has been much more extensive, but is not without its own set of assumptions (91,92). The wide application of temperature studies to surfactant micelles and bilayers discussed above suggests that temperature - induced microemulsion formation may be amenable to study by FT-IR.

Interfacial Phenomena

The application of FT-IR to a wide variety of interfacial phenomena is possible using either attenuated total reflectance (ATR) or grazing angle reflection -absorption. The details of these optical techniques are discussed in a subsequent chapter, in the context of characterization of Langmuir - Blodgett films and self-assembled mono and multilayers. However, ATR has been applied in a number of other areas which will be mentioned here.

In - situ Studies of Solid-Liquid Interfaces

Protein Adsorption. The development of medical implant polymers has stimulated interest in the use of ATR techniques for monitoring the kinetics of adsorption of proteins involved in thrombogenesis onto polymer surfaces. Such studies employ optical accessories in which an aqueous protein solution (93) or even ex - vivo whole blood (94-96) can be flowed over the surface of the internal reflection element (IRE), which may be coated with a thin layer of the experimental polymer. Modern FT-IR spectrometers are rapid - scanning devices, and hence spectra of the protein layer adsorbed onto the IRE can be computed from a series of interferograms recorded continuously in time, yielding an effective time resolution of as little as 0.8 s early in the kinetic runs. Such capability is important because of the rapid changes in the composition of the adsorbed protein layers which can occur in the first several minutes (97).

These kinetics studies required development of reproducible criteria of subtraction of the H-O-H bending band of water, which completely overlaps the Amide I (1650 cm^{-1}) and Amide II (1550 cm^{-1}) bands (98). In addition, correction of the kinetic spectra of adsorbed protein layers for the presence of "bulk" unadsorbed protein was described (99). Examination of kinetic spectra from an experiment involving a mixture of fibrinogen and albumin showed that a stable protein layer was formed on the IRE surface, based on the intensity of the Amide II band. Subsequent replacement of adsorbed albumin by fibrinogen followed, as monitored by the intensity ratio of bands near 1300 cm^{-1} (albumin) and 1250 cm^{-1} (fibrinogen) (93). In addition to the total amount of protein present at an interface, the possible perturbation of the secondary structure of the protein upon adsorption is of interest. Deconvolution of the broad Amide I,II, and III bands can provide information about the relative amounts of α helices and β sheet contents of aqueous protein solutions. Perturbation of the secondary structures of several well characterized proteins were correlated with the changes in the deconvoluted spectra. Combining information from the Amide I and III (1250 cm^{-1}) bands is necessary for evaluation of protein secondary structure in solution (100).

The adsorption-induced changes in secondary structure of fibronectin, on polyurethane block copolymers (101), and human serum albumin, on acrylic polymers used in the manufacture of contact lenses (102), included hydrogen bonding of the protein to the polymer surface and a gain in β sheet conformations at the expense of α helices in the adsorbed layers. In - vivo spoilage of contact lenses involves deposition of lysozyme. A cleaning of these lenses with papain enzyme resulted in incomplete lysozyme removal, and some potentially harmful adsorption of papain itself (103). The differences in adsorption of albumin α helices adsorbed on a germanium IRE, and on a layer of adsorbed albumin, were investigated by a comparison of the ATR spectra of Langmuir-Blodgett films of different thicknesses. A rearrangement of the albumin in these films occurred upon storage at 0 °C (104).

ATR studies of the biocorrosion of submerged copper surfaces have been reported. The IRE of a cylindrical internal reflectance cell (CIRCLE) was coated with a thin copper layer via a vacuum deposition technique (105). The copper layer reduces the sampling depth of the radiation outward from the surface of the IRE. Therefore, the intensity of the water bending band will vary with copper layer thicknesses of 4.1 nm or less. The copper layers were shown to be stable to exposure to water alone, but the presence of acidic polysaccahrides in the water caused a reduction in the copper layer thicknesses (106,107). The adsorption of a model compound, Gum Arabic, onto the coated IRE was detected by increases in the C-O stretching band of the pyranose units near 1050 cm^{-1} (106).

Surfactants at Interfaces. Somewhat surprisingly, the successes described above in the in-situ studies of protein adsorption have not inspired extensive applications to the study of the adsorption of surfactants. The common materials used in the fabrication of IREs, thalliumbromoiodide, zinc selenide, germanium and silicon do, in fact, offer quite a range in adsorption substrate properties, and the potential of employing a thin layer of a substance as a modifier of the IRE surface which is presented to a surfactant solution has also been examined in the studies of proteins. Based on the appearance of the studies described below, and recent concerns about the kinetics of formation of self-assembled layers, (108) it seems likely that in-situ ATR studies of small molecules at solid - liquid interfaces ("wet" solids), will continue to expand in scope.

The IRE of a CIRCLE has also been used as a support for thin layers of solid hydrocarbon and triglyceride "model soils" in FT-IR studies of detergency (109-111). Time - resolved spectra of the solid soil layers during interaction with aqueous solutions of nonionic surfactants were obtained using GC-IR data acquisition software. The removal of the solid soils from the surface of the IRE was monitored by the intensity of the CH$_2$ deformation band near 1470 cm^{-1}. The time - resolved spectra indicated the necessity of penetration of surfactant molecules into the hydrocarbon layers, with a consequent disordering of the methylene chain subcell packing, for successful detergency. In the case of solid tristearin as a model soil, the surfactant penetration resulted in selective removal of the α crystal polymorph, and an acceleration of the conversion of the α polymorph to the β polymorph. This solid phase transition of the triglyceride soil molecules was detrimental to the apparent rate of removal from the IRE surface (111).

Quantitative determinations of the thicknesses of a multiple - layered sample (for example, two polymer layers in intimate contact) by ATR spectroscopy has been shown to be possible. The attenuation effect on the evanescent wave by the layer in contact with the IRE surface must be taken into account (112). Extension of this idea of a step-type concentration profile for an adsorbed surfactant layer on an IRE surface was made (113), and equations relating the Gibbs surface excess to the absorbance in the infrared spectrum of a sufficiently thin adsorbed surfactant layer were developed. The addition of a thin layer of a viscous hydrocarbon liquid to the IRE surface was investigated as a model of a liquid-liquid interface (114) for studies of metal extraction (Ni^{+2}, Cu^{+2}) by a hydrophobic chelating agent. The extraction of the metals from an aqueous buffer into the hydrocarbon layer was monitored kinetically by the appearance of bands unique to the complex formed.

An early paper on the use of ATR for the in - situ monitoring of the adsorption of AOT from heptane and water solutions utilized the intensity of the C=O band (near 1740 cm^{-1}) and the sulfonate band (1045 cm^{-1}) for construction of adsorption isotherms. In addition, the utility of detailed examination of the shifts in the sulfonate band frequency for assessing the headgroup environment in adsorbed layers was discussed (115).

The adsorption of sodium oleate on a fluorite (CaF_2) IRE has been quantitatively studied more recently (116). Adsorption proceeded rapidly during the first two hours, but slowed at longer times, taking as much as 30 hours to reach equilibrium. The formation of interfacial calcium dioleate was clearly indicated by the pair of carboxylate bands at 1571 and 1534 cm^{-1} in an in - situ spectrum of an adsorbed layer formed from sodium oleate at 9 x 10^{-5} M. Calculation of the adsorption isotherm from the FT-IR data, at sodium oleate concentrations between 5 x 10^{-7} and 5 x 10^{-4} M, yielded results in good agreement with values obtained by other methods.

In a study aimed at using in - situ ATR for monitoring the kinetics of the reaction of a silica surface with a silane (117), the details of generation of an oxide surface on a silicon IRE were investigated. The effect of the oxide overlayer on the evanescent wave was evaluated spectroscopically. The thickness of the overlayer, calculated from the absorbance of a test solute, agreed fairly well with an independent ellipsometric determination. Adsorption of diphenylchlorosilane onto the oxide surface occurred without reaction initially, as indicated by the unperturbed Si-H band near 2175 cm^{-1}. Reaction of the silane occurred later, with an accelerating rate between 30 and 60 minutes, and completion of the reaction within 90 minutes. The total amount of silane at the oxide surface, estimated from the aromatic C-H band intensity near 3070 cm^{-1}, appeared constant after 60 minutes, suggesting changes in the nature of the adsorbed layer at later stages of the surface reaction.

Polymer Adsorption. A review of the theory and measurement of polymer adsorption points out succinctly the distinquishing features of the behavior of macromolecules at solid - liquid interfaces (118). Polymer adsorption and desorption kinetics are more complex than those of small molecules, mainly because of the lower diffusion rates of polymer chains in solution and the "rearrangement" of adsorbed chains on a solid surface, characterized by slowly formed, multi-point attachments. The latter point is one which is of special interest in protein adsorption from aqueous solutions. In the case of proteins, initial adsorption kinetics may be quite rapid. However, the slow rearrangement step may be much more important in terms of the function of the adsorbed layer in natural processes, such as thrombogenesis or biocorrosion / biofouling caused by cell adhesion.

The polydispersity of polymers results in competing adsorption of the thermodynamically favored larger molecules for surface sites filled initially by smaller molecules. Different segments of a block copolymer may exhibit quite different adsorption characteristics, complicating the rearrangement process further. This is an effect of considerable interest in protein adsorption, and is referred to as the rearrangement of a protein layer to maximize hydrophobic interaction of "oily" patches with low energy surfaces such as medical implant polymers.

Several IR studies of polymer adsorption are mentioned in reference 118, most of which employed transmission optics to record the spectra of small particles bearing an adsorbed layer.

An example of the use of a CIRCLE for in-situ kinetic studies of polymer adsorption on a germanium surface has recently appeared (119). Mixtures of the protio and deutero forms of polymethylmethacrylate were used. C-H and C-D stretching bands unique to the two polymers, which differed in molecular weight, were thus available for monitoring the composition of the layers adsorbed from CCl$_4$ solution. The C=O band of the ester groups could be used as an indication of total adsorbed polymer. Displacement of an adsorbed layer of lower molecular weight polymer by a dilute solution of a higher molecular weight polymer was detected.

The application of grazing angle reflectance spectroscopic techniques (GRAS) to high polymer adsorption has not been as extensive as in the area of self-assembled monolayers. However, the adsorption of ethylene sulfide modified polystyrene on gold surfaces from toluene was demonstrated by GRAS of dried surfaces (120).

Addition of a thin-film heater to the back of a metal sample holder allowed studies of thin polymer films by GRAS at up to 200 °C (121). In a study of an ultrathin film of polymethylmethacrylate, the two doublets near 1240/1270 and 1150/1190 cm^{-1} exhibited changes in relative intensity above the glass transition temperature of 100 °C, indicating that the polymer glass structure was maintained even in such thin films.

Air-Water Interface. Organized films of surfactants and phospholipids at the air-water interface are of interest in biophysics, general interfacial chemistry, and have relevance to the self-assembling aggregates, which are viewed as having potential applications in non-linear optics and as microelectronic devices (122). FT-IR spectroscopy has recently been applied to the problem of obtaining information about amphiphiles at the air-water interface.

Using the surface of liquid water as the reflective substrate, infrared reflection-absorption spectra of phospholipid layers have been obtained (123-125). The infrared beam is taken external to the spectrometer with two CaF$_2$ lenses and a flat mirror, with another flat mirror directing the reflected beam to the detector. A Langmuir film balance was designed (124) which allowed compression of the phospholipid layers to over 50 dyne/cm.

The frequencies of the C-H stretching bands in the spectra of distearoyl- and dimyristoylphosphatidylcholine films on water indicate differences in disorder of the methylene chains between liquid - condensed and liquid - expanded films. In the condensed films, the methylene chains are frozen in an all - trans configuration, while those of the expanded films are highly disordered. The important headgroup bands (phosphate P=O and choline C-N stretching) between 1300 and 900 cm^{-1} are also detected. The capability to simultaneously monitor headgroup and tail environments is an important one, as has been mentioned above in the studies of lipid phase behavior, since interactions of the headgroups with the subphase of surface films could be investigated.

Unlike grazing angle reflection studies of films on metals, both parallel and perpendicularly polarized radiation is observable upon reflection from the water surface. Similar to ATR experiments, electric fields in all three dimensions relative to the water surface are present, indicating the potential for complete orientation studies of the surface layers(126).

Concluding Remarks

There are several major areas of interfacial phenomena to which infrared spectroscopy has been applied that are not treated extensively in this volume. Most of these areas have established bodies of literature of their own. In many of these areas, the replacement of dispersive spectrometers by FT instruments has resulted in continued improvement in sensitivity, and in the interpretation of phenomena at the molecular level. Among these areas are: the characterization of polymer surfaces with ATR (127-129) and diffuse reflectance (130) sampling techniques; transmission IR studies of the surfaces of powdered samples with adsorbed gases (131-136); alumina(137,138), silica (139), and catalyst (140) surfaces; diffuse reflectance studies of organo- modified mineral and glass fiber surfaces (141-143); metal overlayer enhanced ATR (144) ; and spectroelectrochemistry (145-149).

This review obviously reflects the preferences of the author in stressing the application of FT - IR to amphiphile phase behavior and the kinetics of interfacial phenomena. In some sense, FT - IR may be considered an emerging technique in studies of the phase behavior of surfactants. However, from the wide range of studies of lipid bilayers mentioned, it seems that the concept of using FT - IR to probe surfactant molecule associative properties in other aggregates such as micelles, gels and vesicles can be considered a logical extension.

In the area of interfacial chemistry, it is quite likely that FT - IR techniques will continue to provide rapid, non - destructive analyses of self - assembled layers, perhaps with extended work on the kinetics of the formation of these potentially useful aggregates. In - situ kinetics studies of surfactant, polymer and protein adsorption hold great potential for improving our understanding of processes as diverse as detergency, the rejection of medical implant devices, or perhaps the interaction of a valuable genetically engineered enzyme with a fermentation vessel or transport tubing.

Literature Cited

1. Griffiths, P. R.; de Haseth, J. A. *Fourier Transform Infrared Spectrometry*; Wiley-Interscience: New York, NY; 1986
2. Griffiths, P. R. *Chemical Infrared Fourier Transform Spectroscopy*; Wiley - Interscience: New York, NY; 1975
3. *Fourier Transform Infrared Spectroscopy: Applications to Chemical Systems*; Ferraro, J. R.; Basile, L. J.; Eds.; Academic Press: New York, NY 1979; Vol. 2
4. Cooley, J. W.; Tukey, J. W. *Math. Comput.* **1965**, 19, 297
5. Michelson, A. A. *Light Waves and Their Uses*; Pheonix Science Series; University of Chicago Press: Chicago, IL, 1962
6. Felgett, P. In *Aspen Intl. Conf. on Fourier Spect.*; Vanasse, G. A., Stair, A. T., Baker, D. J., Eds.; AFCRL - 71 - 0019 , 1970, 139
7. Cameron, D. G.; Kauppinen, J. K.; Moffatt, D. J.; Mantsch, H.H. *Applied Spectroscopy* **1982**, 36, 245
8. Cameron, D. G.; Kauppinen, J. K.; Moffatt, D. J.; Mantsch, H. H. *Appl. Spec.* **1981**, 35, 271
9. Cameron, D. G.; Kauppinen, J. K.; Moffatt, D. J.; Mantsch, H. H. *Appl. Opt.* **1981**, 20,1986
10. Cameron, D. G.; Kauppinen, J. K.; Moffatt, D. J.; Mantsch, H. H. *Anal. Chem.* **1981**, 53,1454
11. Kauppinen, J. K. In *Spectrometric Techniques*; Vanasse, G. A., Ed.; Academic Press: New York, NY, 1983, Vol III;pp 199
12. Griffiths. P. R.; Yang, W. *Computer Enhanced Spec.* **1984**, 1,157
13. Griffiths, P. R.; Pariente, G. L. *Trends in Anal. Chem.* **1986**, 5, 209
14. James, D. I.; Maddams, W. F. *Appl. Spec.* **1987**, 41, 1362
15. Maddams, W. F. *Appl. Spec.* **1980**, 34, 245
16. Antoon, M. K.; D'Esposito, L.; Koenig, J. L. *Appl. Spec.* **1979**, 33, 351
17. Koenig, J. L.; Tovar Rodriquez, M. *Appl. Spec.* **1981**, 35, 543
18. Savitsky, A.; Golay, M. J. E. *Anal. Chem.* **1964**, 36, 1627
19. Holler, F.; Callis, J. B. *J. Phys. Chem.* **1989**, 93, 2053
20. Holler, F.; Callis, J. B.; Burns, D. H. *Appl. Spec.* **1989,** 43, 877
21. Mantsch, H.H.; Moffatt, D. J.; Casal, H. L. *J. Mol. Struct.* **1988**, 173, 285
22. Chapman, D.; Gomez-Fernandez, J. C.; Goñi, F. M.; Barnard, M. J. *Biochem. Biophys. Methods* **1980**, 2, 315
23. Umemura, J.; Cameron, D. G.; Mantsch, H.H. *J. Phys. Chem.* **1980**, 84, 2272
24. Rein, A. J.; Wilks, P. *Am. Lab.* **1982**, 14,152
25. Harrick, N. J. *Appl. Spec.* **1983**, 37, 573
26. Chittur, K. K.; Fink. D. J.; Leininger, R. I.; Hutson, T. B. *J. Coll. Int. Sci.* **1986**, 111, 419
27. Israelachvili, J. N.; Mitchell, D. J.; Ninham, B. W. *J. Chem. Soc. Faraday Trans.* **1976**, 72, 1525
28. Kaler, E. W.; Murthy, A. K.; Rodriquez, B. E.; Zasadzinsky, J. A. N. *Science* **1989** 245, 1371
29. Maroncelli, M.; Qi, S. P.; Strauss, H. L.;Snyder, R. G. *J. Am. Chem. Soc.* **1982**, 104, 6237

30. Maroncelli, M.;Qi, S. P.; Strauss, H. L.; Snyder, R. G. *Science* **1981**, 214, 188
31. Snyder, R. G.; Schachtschneider, J. H. *Spectrochim. Acta* **1963**, 19, 85
32. Snyder, R. G.; Schachtschneider, J. H. *Spectrochim. Acta* **1963**, 19, 117
33. Snyder, R. G.; Schachtschneider, J. H. *Spectrochim. Acta* **1964**, 20, 853
34. Snyder, R. G. *J. Mol. Spectry.* **1967**, 23,224
35. Snyder, R. G. *J. Chem. Phys.* **1967**, 47, 1316
36. Casal, H. L.; Mantsch, H. H.; Cameron, D. G.; Snyder, R. G. *J. Chem. Phys.* **1982**, 77,2825
37. Snyder. R. G. *J. Chem. Phys.* **1979**, 71, 3229
38. Snyder, R. G.; MacPhail, R. A.; Strauss, H. L.; *J. Am. Chem. Soc.* **1980**, 102, 3976
39. Maroncelli, M.; Strauss, H. L.; Snyder, R. G.; Elliger, C. A.; Casal, H. L.; Mantsch, H. H.; Cameron, D. G.; *J. Am. Soc.* **1983**, 105, 134
40. Holland, R. F.; Nielsen, J. R.; *J. Mol. Spec.* **1962**, 8, 383
41. Casal, H. L.; Mantsch, H. H. *Biochimica et Biophysica Acta* **1984**, 779, 381
42. Cameron, D. G.; Mantsch, H. H.; *Biophys. J.* **1982**, 38, 175
43. Cameron, D. G.; Mantsch, H. H.; Tremblay, P. A.; Kates, M. *Biochim. Biophys. Acta* **1982**, 689, 63
44. Cameron, D. G.; Mantsch, H. H.; Umemura, J. *Biochim. Biophys. Acta* **1980**, 602, 32
45. Fookson, J. E.; Wallach, D. F. H. *Archives of Biochemistry and Biophysics* **1978**, 189, 195
46. Fringeli, U. P.; Gunthard, H. In *Membrane Spectroscopy*; Grell, E.; Ed.; Springer-Verlag: New York, NY, 1981; p. 270
47. Dluhy, R.A.; Cameron, D. G.; Mantsch, H. H.; Mendelsohn, R. *Biochemistry* **1983**, 22, 6318
48. Cameron, D. G.; Gudgin, E.; Mantsch, H. H. *Biochemistry* **1981**, 20, 4496
49. Cameron, D. G.; Mantsch, H. H.; Casal, H. L.; Boulanger, Y.; Smith, I. C. P. *Biophys. J.* **1981**, 35, 1
50. Dluhy, R. A.; Moffatt, D.J.; Cameron, D. G.; Mendelsohn, R.; Mantsch, H. H. *Can. J. Chem.* **1985**, 63, 1925
51. Umemura, J.; Mantsch, H. H.; Cameron, D. G. *J. Coll. Int. Sci.* **1981**, 83, 558
52. Jaworsky, M.; Mendelsohn, R. *Biochemistry* **1985**, 24, 3422
53. Mendelsohn, R.; Anderle, G.; Jaworsky, M.; Mantsch, H. H.; Dluhy, R. A. *Biochim. Biophys. Acta* **1984**, 775, 237
54. Mendelsohn, R.; Brauner, J. W.; Faines, L.; Mantsch, H. H.; Dluhy, R. A. *Biochim. Biophys. Acta* **1983**, 774, 237
55. Dluhy, R. A.; Mendelsohn, R.; Casal, H. L.; Mantsch, H. H. *Biochemistry* **1983**, 22, 1170
56. Surewicz, W.; Moscarello, M. A.; Mantsch, H. H. *Biochemistry* **1987**, 26, 3881
57. Simada, I.; Ishida, H.; Ishitani, A.; Kunitake, T. *J. Coll. Int. Sci.* **1987**, 120, 523
58. Kawai, T.; Umemura, J.; Takenaka, T.; Gotou, M.; Sunamoto, J. *Langmuir* **1988**, 4, 449
59. Bunding Lee, K. A. *J. Phys. Chem.* **1989**, 93, 926
60. Nakashima, N.; Yamada, N.; Kunitake, T.; Umemura, J.; Takenaka, T. *J. Phys. Chem.* **1986**, 90, 3374
61. Kawai, T.; Umemura, J.; Takenaka, T.; Kodama, M.; Seki, S. *J. Coll. Int. Sci.* **1985**, 103, 56
62. Umemura, J.; Kawai, T.; Takenaka, T. *Mol. Cry. Liq. Cry.* **1984**, 112, 293
63. Kawai, T.; Umemura, J.; TAkenaka, T.; Kodama, M.; Ogawa, Y.; Seki, S. *Langmuir* **1986**, 2, 739
64. Evans, D. F.; Warr, G. G. *Langmuir* **1988**, 4, 217
65. Wunderlich, I.; Hoffmann, H.; Rehage, H. *Rheol. Acta* **1987**, 26, 532
66. Mantsch, H. H.; Kartha, V.B.; Cameron, D. G. In *Surfactants in Solution*; Lindman, B.; Mittal, K., Eds.; Plenum Press: New York, NY, 1984, Vol. 1

67. Cameron, D. G.; Umemura, J.; Wong, P. T. T.; Mantsch, H. H. *Colloids and Surfaces* **1982**, 4, 131
68. Mantsch, H. H.; Yang, P. W. *J. Coll. Int. Sci.* **1986**, 113, 218
69. Cameron, D. G.; Charette, G. M. *Appl. Spec.* **1981**, 35, 224
70. Hayakawa, S.; Matsui, T.; Tanaka, S. *Appl. Spec.* **1984**,41, 1438
71. Villalain, J.; Gomez-Fernandez, J.; Prieto, M. *J. J. Coll. Int. Sci.* **1988**, 124, 233
72. Casal, H. L. *J. Am. Chem. Soc.* **1988**, 110,5203
73. Bellamy, L. J. *The Infrared Spectra of Complex Molecules*; Chapman and Hall: London 1975; Third Edition
74. Wong, P. T. T.;Siminovitch, D. J.; Mantsch, H. H. *Bichom. Biophys. Acta* **1988**, 947, 139
75. Wong, P. T. T.; Mantsch, H. H. *J. Coll. Int. Sci.* **1989**, 129, 258
76. Wong, P. T. T.; Casal, H. L.; *J. Phys. Chem.* **1990**, 94, 777
77. *Phenomena in Mixed Surfactant Systems*; Scamehorn, J. F.; Ed.; ACS Symposium Series 311; American Chemical Society: Washington, D.C., 1986
78. Scheuing, D. R.; Weers, J. G. *Langmuir* **1990**, 6, 665
79. Scheuing, D. R.; Weers, J. G.; Rathman, J. F. *Colloid and Polymer Science* ; in press
80. Fendler, E. J.; Chang, S.; Fendler, J. H.; Madery, R. T.; El Seoud, O. A.; Woods, V. A.; In *Reaction Kinetics in Micelles;* Cordes, E. H., Ed.; Plenum Press: New York, NY, 1973 p. 127
81. Luisi, P.L.; Magid, L.J. *CRC Crit. Rev. Biochem.* **1986**, 20, 409
82. Walde, P.; Luisi, P.L. *Biochemistry* **1989**, 28, 3353
83. Kise, H.; Iwamoto, K.; Seno, M. *Bull. Chem. Soc. Jpn.*, **1982**, 55, 3856
84. Wu, J.; Shi, N.; Xu, X.; Xu, G. *Infrared Physics* **1984**, 24, 245
85. Rosenholm, J.; Sjöblom, J.; Österholm, J. *Chem. and Phys. of Lipids,* **1982**, 31, 117
86. MacDonald, H.; Bedwell, B.; Gulari, E. *Langmuir* **1986**, 2, 704
87.Boicelli, C.; Giomini, M.; Giuliani, A. *Appl. Spec.* **1984**, 38, 537
88. Nagy, J.; Yuan, Y.; Jao, T.; Fendler, J. H. *J. Phys. Chem.* **1990**, 94, 863
89. Smith, R. D.; Fulton, J. L.; Blitz, J. P.; Tingey, J. T. *J. Phys. Chem.* **1990**, 94, 781
90. Pacynko, W. F.; Yarwood, J.; Tiddy, G. J. T. *J. CHem. Soc. Faraday Trans.* **1989**, 85, 1397
91. Shinoda, K.; Lindman, B. *Langmuir* **1987**, 3, 135
92. Lindman, B.; Söderman, O.; Wennerström, H.; In *Surfactant Solutions: New Methods of Investigation*; Zana, R.; Ed.; Marcel Dekker: New York, NY, 1986
93. Gendreau, R. M.; Leininger, R. I.; Winters, S.; Jakobsen, R. J. In *Biomaterials: Interfacial Phenomena and Applications;* Cooper, S. L.; Peppas, N. A.; Eds.; and Hoffman, A.S.; Ratner, B. D.; Assoc. Eds.; Advances in Chemistry Series 199; American Chemical Society: Washington, D. C., 1982
94. Gendreau, R. M.; Leininger, R. I.; Winters, S.; Jakobsen, R. J.; Fink, D.; Hassler, C. R. *Appl. Spec.* **1981**, 35, 353
95. Gendreau, R. M.; Leininger, R. I.; Winters, S.; Jakobsen, R. J. *Appl. Spec.* **1982**, 36, 404
96. Gendreau, R. M.; Jakobsen, R. J. *J. Biomaterials Research* **1979**, 13, 893
97. Chittur, K. K.; Fink, D. J.; Leininger, R. I.; Hutson, T. B. *J. Coll. Int. Sci.* **1986**, 111, 419
98. Powell, J. R.; Wasacz, F. M.; Jakobsen, R. J. *Appl. Spec.* **1986**, 40, 339
99. Chittur, K. K.; Fink, D. J.; Leininger, R. I.; Hutson, T. B.; Gendreau, R. M.; Jakobsen, R. In *Proteins at Interfaces: Physiochemical and Biochemical Studies*; Brash, J. L.; Horbett, T. A.; Eds.; ACS Symposium Series 343; American Chemical Society: Washington, D.C., 1987; Chap. 23
100. Jakobsen, R. J.; Wasacz, F. M.; In *Proteins at Interfaces: Physiochemical and Biochemical Studies*; Brash, J. L.; Horbett, T. A.; Eds.; ACS Symposium Series 343;

American Chemical Society: Washington, D.C., 1987; Chap. 22.
101. Pitt, W. G.; Spiegelberg, S. H.; Cooper, S. L.;In *Proteins at Interfaces: Physiochemical and Biochemical Studies*; Brash, J. L.; Horbett, T. A.; Eds.; ACS Symposium Series 343; American Chemical Society: Washington, D.C., 1987; Chap. 21
102. Castillo, E. J.; Koenig, J. L.; Anderson, J. M.; Lo, J. *Biomaterials* **1984**, 5, 311
103. Castillo, E. J.; Koenig, J. L.; Anderson, J. M. *Biomaterials* **1986**, 7, 89
104. Jakobsen, R. J.; Cornell, D. G. *Appl. Spec.* **1986**, 40, 318
105. Iwaoka, T.; Griffiths, P. R.; Kitasako, J. T.; Geesey, G. G. *Appl. Spec.* **1986**, 40, 1062
106. Iwaoka, T.; Griffiths, P. R.; Geesey, G. G. *J. Coll. Int. Sci.* **1987**, 120, 370
107. Jolley, J. G.; Geesey, G. G.; Hankins, M. R.; Wright, R. B.; Wichlacz, P. L. *Appl. Spec.* **1989**, 43, 1062
108. Chen, S. H.; Frank, C. W. *Langmuir* **1989**, 5, 978
109. Scheuing, D. R. *Appl. Spec.* **1987**, 41, 1343
110. Scheuing, D. R.; Hsieh, J. C. L. *Langmuir* **1988**, 4, 1277
111. Scheuing, D. R. *Langmuir* **1990**, 6, 312
112. Ohta, K.; Iwamoto, R. *Appl. Spec.* **1985**, 39, 418
113. Sperline, R. P.; Muralidharan, S.; Freiser, H. *Langmuir* **1987**, 3, 198
114. Sperline, R. P.; Freiser, H. *Langmuir* **1990**, 6, 344
115. McKeigue, K.; Gulari, E.; In *Surfactants in Solution*; Mittal, K.; Lindman, B.; Eds.; Plenum Press: New York, NY, 1984, Vol. 2
116. Keller, J. J.; Cross, W. M.; Miller, J. D. *Appl. Spec.* **1989**, 43, 8
117. Parry, D.B.; Harris, J. M. *Appl. Spec.* **1988**, 42, 997
118. Vincent, B.; Whittington, S. G.; In *Surface and Colloid Science*; Matijevic, E., Ed.; Plenum Press: New York, NY, 1982, Vol. 12
119. Kuzmenka, D. J.; Granick, S. *Colloids and Surfaces* **1988**, 31, 105
120. Stouffer, J. M.; McCarthy, T. J. *Polymer Preprints* **1985**, 27 (2), 242
121. Rabolt, J. F.; Schlotter, N. E., *Appl. Spec.* **1985**, 39, 994
122. Swalen, J. D.; Allara. D.L.; Andrade, J.D.; Chandross, E. A.; Garoff, J.; Israelachvilli, J.; McCarthy, T. J.; Murray, R.; Pease, R. F.; Rabolt, J. F.; Wynne, K. J.; Yu, H. *Langmuir* **1987**, 3, 932
123. Dluhy, R. A.; Wright, N. A.; Griffiths, P. R. *Appl. Spec.* **1988**, 42, 138
124. Dluhy, R. A.; Mitchell, M. L.; Pettenski, T.; Beers, J. *Appl. Spec.* **1988**, 42, 1289
125. Dluhy, R. A.; Cornell, D. G.; *J. Phys. Chem.* **1985**, 89, 3195
126. Dluhy, R. A. *J. Phys. Chem.* **1986**, 90,1373
127. Koenig, J. L. *Appl. Spec.* **1975**, 29, 293
128. Coleman, M. M.; Painter, P. C. *J. Macromol. Sci. Rev. Macromol. Chem.* C16, **1977**, 197
129. Sung, C. S. P. *Macromolecules* **1981**, 14, 591
130. Fuller, M. P.; Griffiths, P. R. *Anal. Chem.* **1978**, 50, 1906
131. Hair, M. L. *Infrared Spectroscopy in Surface Chemistry*; Marcel Dekker: New York, NY, 1967
132. *Vibrational Spectroscopies for Adsorbed Species;* Bell, A. T.; Hair, M. L.; Eds.; ACS Symposium Series 137; American Chemical Society: Washington, D.C., 1980
133. Kiselev, A. V.; Lygin, V. I.; *Infrared Spectra of Surface Compounds;* Wiley: New York, NY, 1978
134. Little, L. H. *Infrared Spectra of Adsorbed Species;* Academic Press: New York, NY, 1966
135. Gardella Jr., J.; Jiang, D.; Eyring, E. M. *Appl. Spec.* **1983**, 37, 131
136. Martin, M.; Childers, J. W.; Palmer, R. A. *Appl. Spec.* **1987**, 41, 20
137. Chen, J. G.; Basu, P.; Ballinger, T. H.; Yates, J. T. *Langmuir* **1989**, 5, 352
138. Basu, P.; Ballinger, T. H.; Yates, J. T. *Langmuir* **1989**, 5, 502

139. Severdia, A. G.; Low, M. J. D. *Langmuir* **1988**, 4, 1234
140. Zhou, X.; Gulari, E. *Langmuir* **1988**, 4, 1332
141. Vagberg, L.; DePotocki, P.; Stenius, P. *Appl. Spec.* **1989**, 43, 1240
142. Urban, M.; Koenig, J. L. *Appl. Spec.* **1986**, 40, 513
143. Sivamohan, R.; deDonato, P.; Cases, J. M. *Langmuir* **1990**, 6, 637
144. Ishida, H.; Ishino, Y.; *Appl. Spec.* **1988**, 42, 1296
145. Nazri, G.; Corrigan, D. A.; Maheswari, S.; *Langmuir* **1989**, 5, 17
146. Mu, X. H.; Kadish, K. M. *Langmuir* **1990**, 6, 51
147. Guillaume, F.; Griffin, G.L.; *Langmuir* **1989**, 5, 783
148. Harris, J. M.; Parry, D. B.; Ashley, K. *Langmuir* 1990, 6, 209
149. Bunding Lee, K. A. *Langmuir* **1990**, 6, 709

RECEIVED August 2, 1990

COLLOIDAL AGGREGATES

Chapter 2

CD_2 Rocking Modes as Quantitative Fourier Transform Infrared Spectroscopic Probes of Conformational Disorder in Phospholipid Bilayers

Richard Mendelsohn and Mark Allen Davies

Department of Chemistry, Newark College of Arts and Science, Rutgers University, 73 Warren Street, Newark, NJ 07102

The quantitative determination of phospholipid acyl chain conformational disorder in hydrated systems of dipalmitoylphosphatidylcholine (DPPC), dipalmitoylphosphatidylethanolamine (DPPE), mixtures containing DPPC and cholesterol, and mixtures of DPPC and Gramicidin D has been made using FT-IR. The method takes advantage of the sensitivity of CD_2 rocking modes to the trans-gauche rotational isomerization process. The relative amounts of trans and gauche conformers in the above systems have been monitored as a function of temperature and position of the CD_2 marker along the acyl chain (bilayer depth). At 49°C, above the main gel to liquid crystal transition of DPPC, the percentages of gauche conformers present are 20.7±4.2, 32.3±2.3, and 19.7±0.8 for DPPC derivatives deuterated at the 4, 6, and 10 acyl chain positions, respectively. In 2:1 DPPC/cholesterol mixtures, gauche rotamer formation is inhibited by factors of ~9 and ~6 at the 6 and 4 positions, respectively. A large decrease, from 78% to 15%, in the percentage of single gauche bend or kink conformers relative to multiple gauche forms was observed in Gramicidin D containing mixtures as the lipid/peptide mole ratio was decreased from 30:1 to 10:1. It was observed that both DPPC and DPPE exhibit a curiously large amount of disorder at the 4 position in the gel phase.

0097–6156/91/0447–0024$06.00/0

The fluid mosaic model, as proposed nearly two decades ago by Singer and Nicolson (1), forms the current conceptual framework for biophysical investigations of biological membrane structure. Within the context of the model, it has been recognized that the lipid bilayer in native membranes behaves as a two-dimensional fluid whose conformational and motional properties are controlled by its composition. It is also generally assumed that the structural and dynamic organization of the bilayer is geared toward optimizing the interactions and functions of membrane proteins.

In view of the often demonstrated (2,3 and references therein) importance of bilayer organization/structure/fluidity in controlling membrane function, numerous biophysical studies have appeared over the last two decades attempting to detail, at the molecular and supramolecular levels, the structural and dynamic properties of biological membrane models. Methods such as ESR, ^{2}H NMR, fluorescence, Raman and IR spectroscopies have all been useful in demonstrating the existence of a variety of characteristic phospholipid individual and collective motions. However, a quantitative picture of the contribution of any particular motion to phospholipid dynamics or ordering has remained elusive. Two reasons for this deficiency are apparent. First, two of the techniques, fluorescence and ESR spectroscopies, require probe molecules of such bulkiness and complexity that it becomes unclear how to extrapolate from the measured spectral properties of the probe to the native system. In addition, the environments and dynamics reported upon may be substantially altered by the bulky reporter groups (4). However, the substantial sensitivity of these two methods has resulted in important, but essentially qualitative, information becoming available from systems as complex as native membranes (5,6).

The second reason why complete dynamic and conformational information is not available is related to the characteristic time scale of particular experiments. For example, the elegant techniques of ^{2}H NMR spectroscopy have been widely applied to biological membranes (7-9, and references containied therein). The measured quadrupolar splittings lead directly to order parameters that are averaged by all motions faster than about a microsecond, the characteristic time scale for this measurement. Elaborate, usually model dependent, calculations are required to ascertain the contributions of any one of the possible acyl chain motions (trans-gauche isomerization, chain librations, etc.) to the measured spectral parameters (10,11).

The vibrational spectroscopy time scale (10-300 x 10^{-15} sec) is appropriate for the direct sampling of the fastest motion expected to occur in phospholipid acyl chains, namely trans-gauche isomerization. To date, most Raman and FT-IR studies of phospholipid phase behavior and lipid/protein interaction have focused mainly on qualitative measures of acyl chain organization. For example, the CH_2 symmetric stretching modes near 2850 cm^{-1} undergo a frequency increase of 2-5 cm^{-1} during phospholipid gel→liquid crystal phase transitions (12 and references contained therein). However, the gel and liquid crystal phase spectra are highly overlapped and are not easily resolvable in a unique fashion. Thus, quantitative

determination of the extent of disorder induced during the
transition is difficult, although attempts from this laboratory and
the National Research Council of Canada using two-state models have
been reported ([13,14]). Some success has also been achieved with the
acyl chain 1130 cm^{-1} skeletal optical modes (C-C stretching mode) in
the Raman spectrum. The vibration involved arises primarily from
trans C-C segments in the acyl chains. Gaber and Peticolas ([15])
proposed that each trans C-C bond in a sequence of three makes a
constant intensity contribution to the observed band, and developed
a "trans" Raman order parameter. However , the assumption of
constant intensity contribution per trans bond has been challenged
by Pink et al. ([16]). The latter have elaborated a theory of chain
conformations as a function of temperature and have discussed the
significance of the 1130 cm^{-1} intensities within the context of
their theory. The consistency between theory and experiment is good
for bilayers of saturated phosphatidylcholines, although the
theoretical framework for the intensity calculation was challenged
by Vogel and Jahnig ([17]). In any case, it is not evident how to apply
the model in biologically significant situations, where there is no
applicable theory of conformations. Overall, the quantitative use of
this mode is hampered by ambiguity as to the relationship between
its intensity and the number and location of trans bonds in an acyl
chain.
      An additional vibrational mode used in attempts to
quantitatively characterize acyl chain order is the longitudinal
acoustical "accordion" mode in the Raman spectrum. The frequency of
this low frequency bending motion has been found to vary inversely
with the all-trans acyl chain length ([17,18]). Ordered lipid phases
are found to possess longitudinal acoustical frequencies appropriate
for their chain length. The band broadens in the liquid crystal
phase, suggesting a distribution of chain lengths in residual
all-trans segments. No useful model for the intensity of this band
in fluid phases has been presented.
      The problem of finding those vibrational modes of hydrocarbon
chains that relate directly to conformational structure has been
extensively addressed by Snyder and his colleagues ([19-21]). Snyder
and Poore observed ([19]) in 1973 that the rocking mode vibration of
an isolated CD_2 group is reasonably well localised, appears in a
spectral region free from interference from other acyl chain
vibrations ,and has a frequency which is sensitive on the
conformation of the C-C bonds adjacent to it. Trans-trans,
trans-gauche, and gauche-gauche bond pairs have well separated CD_2
rocking modes. The form of the CD_2 rocking mode in a trans
polyethylene chain is shown in Figure 1. The theoretical framework
for use of these modes in the current application follows.

THEORY

The notation of Maroncelli et al. ([20,21]) will be followed in
discussing conformation, so that, in our designation of the
conformational state of a CD_2 group, the CC bonds that are directly
bonded to the CD_2 group will be underlined. Thus gt means the CD_2
moiety connects a trans and a gauche bond. End-gauche conformations

(an end bond adjoining a gauche bond) are denoted by eg . From normal-coordinate calculations, IR bands observed near 620 and 650 cm^{-1} were assigned to CD_2 rocking modes of bonds in tt and gt sequences. The former is the much stronger feature in the spectra of solid alkanes. The gt band was experimentally found (22) to be split into a doublet with components at 652,646 cm^{-1} (for an interior CD_2) and 657,650 cm^{-1} for a CD_2 group at the penultimate carbon. An assignment of the CD_2 rocking modes consistent with normal coordinate calculations is shown in Table I.

Table I : CD_2 Rocking Modes and Acyl Chain Structure
(Maroncelli, Strauss, and Snyder, J. Phys. Chem. 89, 4390 (1985))

Conformer Class	Conformation	ν_{obsd} (cm^{-1})
et	(end-trans)	622
egt	(end-gauche)	657
egg	(end-gauche)	650
tttt	(trans)	622
ttgtt + tg'tgt	(single gauche bend + kink)	652
ttggt + tg'tggt	(multiple gauche)	646

The sensitivity of the mode to conformation about the C-CD_2-C segment was explained by the coupling of the rocking coordinate to adjacent CH_2 rocking coordinates. The magnitude of the coupling as expressed in the L matrix (Table II) is conformation dependent and does not significantly extend more than one CH_2 group beyond the central unit. The origin of the gauche doublet has been determined (20,21). The high frequency component of the doublet at 652 cm^{-1} has been assigned to conformations termed tgt , encompassing single gauche rotamers (ttgtt) and kinks (tg'tgt) , while the 646 cm^{-1} component arises from double gauche (ttggt) and multiple gauche (tg'tggt) containing forms.

Table II : Normal Coordinate Analysis of CD_2 Rocking Modes
(Maroncelli, Strauss, and Snyder, J. Phys. Chem. 89, 4390 (1985))

Class	ν	L Matrix Elements				
		P_{-2}	P_{-1}	P_{CD2}	P_{+1}	P_{+2}
tttt	622	.015	.202	.487	.202	.015
ttgtt	652	.070	.321	.418	-.052	-.031
tg'tgt	652	-.080	.292	.432	-.067	-.034
ttggt	646	.063	.306	.415	-.112	.155
tg'tggt	646	-.062	.277	.427	-.127	.157

Quantitative determination of conformational disorder, in addition to assignments of the spectral features, requires accurate calibrations of relative absorptivities (extinction coefficients) for the various bands. These were calculated with the aid of the rotational isomeric state model of Flory (22), a statistical mechanical determination of conformational disorder in alkanes. The model requires an estimate for the energies of gauche bonds ,chosen from the best available experiments to be 508 +/- 50 cal/mole, and an energy for the bond pairs E_{gg}, chosen to be 3000 cal/mole. The absorptivity ratios, as defined below, were found to be:

$$A_r(eg) = \frac{I(657) + I(650)}{I(622)} \Big/ \frac{P(eg)}{P(\underline{et})} = 1.14 +/- 0.08 \qquad (1)$$

$$A_r(gt) = \frac{I(652) + I(646)}{I(622)} \Big/ \frac{P(gt)}{P(\underline{tt})} = 1.00 +/- 0.07 \qquad (2)$$

where the P's are the probabilities. The error bars are propagated from the uncertainty in the gauche energies. In addition, the 652 and 646 cm^{-1} components of the gauche doublet are shown to have the same absorptivity, within experimental error. To convert experimental data to fraction gauche rotamers, the following simple relationship thus holds:

$$f_g = \frac{[I(652) + I(646)]}{[I(622) + I(652) + I(646)]} \qquad (3)$$

where f_g is the total fraction of gauche rotamers not at the chain end. The fraction of particular conformational classes corresponding to the 652 and 646 cm^{-1} spectral features, may be obtained by replacement of the numerator of Equation 3 with the appropriate integrated band intensity.

The above theoretical and experimental framework suggests that the CD_2 rocking modes are suitable for determination of conformational disorder in phospholipids. There are, however, substantial technical difficulties which must be overcome prior to a successful experimental outcome. First, specifically deuterated molecules must be synthesized. Relatively high incorporation levels are required since the CHD_1 rocking mode from partially deuterated sites absorbs near 660 cm^{-1} and interferes with the high frequency component of the gauche doublet. Second, the extremely high background of liquid water or better, D_2O, at the measured spectral positions must be overcome. For example, we note that under conditions of our experiment (6 micron cell, D_2O solvent, 1.5:1 (w:w) D_2O/phospholipid ratio), the absorbance of the solvent librations is about 0.4-1.0 A, while the absorbance of the gauche doublet varies between .0001 and .005 A units, values 80-10000 fold reduced from the background.

This chapter summarizes the first direct determinations of conformational disorder in phospholipid bilayer systems. The model phospholipid chosen is 1,2 dipalmitoylphosphatidylcholine (DPPC), whose physical properties have been widely investigated for two

decades,and whose structure is shown in Figure 2.   Four derivatives
of DPPC with CD$_2$ groups at the 4,6,10, and 12 positions in both acyl
chains have been synthesized. The position and temperature
dependence of conformational disorder have been measured in DPPC
multibilayer vesicles (23). In addition,the effects of cholesterol,
an important constituent of mammalian membranes, and Gramicidin D,(a
pore-forming, membrane spanning, antibiotic peptide) on phospholipid
conformational order, have been determined (24,25).

EXPERIMENTAL

PHOSPHOLIPID SYNTHESIS. Phospholipids were synthesized (H.Schuster,
S.S. Hall, and R.Mendelsohn, in preparation) according to the
procedures of Tulloch (26), modified with more modern and efficient
coupling steps, and scaled up to produce 2-3 grams of specifically
deuterated material. Derivatives were fully characterized with NMR,
MS, FT-IR, and Differential Scanning Calorimetry (DSC). Purity is
estimated from NMR data at > 98%. The extent of deuteration, as
estimated from the residual intensity of the CHD rocking modes at
660 cm^{-1} from incompletely deuterated sites,is >95%. Each derivative
showed, by differential scanning calorimetry, a sharp (transition
width 0.3 °C or less) gel→liquid crystal phase transition near 41°C
and a pre-transition between 32-35°C.

SAMPLE PREPARATION. It is stressed that careful attention to detail
is required for the successful execution of this experiment. First,
although a large excess of solvent (D$_2$O) is to be avoided as the
absorbance in the region of interest will be too intense for proper
background subtraction, complete hydration of the phospholipid is
necessary. Incomplete hydration (aside from altering DPPC phase
properties (27)) results in many extra spectral features (of obscure
origin) between 600-700 cm^{-1}, rendering the sought conformational
analysis impossible. Such spurious bands were noted in many
fruitless attempts to monitor this spectral region with evanescent
wave (attenuated total reflectance) FT-IR. The approach was
eventually abandoned in favor of transmission spectroscopy.
    A D$_2$O/lipid ratio of 2:1 (v:w) led to easily handled samples
and consistent spectroscopic results. Typically, 10 mg of
specifically deuterated phospholipid and 20μl D$_2$O were added to a
culture tube, which was then sealed and incubated at 55°C for at
least 1 hr, with periodic agitation to ensure complete dispersal and
hydration. Similar procedures were followed for cholesterol and
Gramicidin D-containing samples. Samples were placed between two
AgCl windows, and sealed with a 6 μM,spacer. The assembly was
wrapped with Teflon tape, as a further seal against dehydration, and
inserted into a Harrick cell. Temperature was controlled to ± 0.1°C
with a Haake circulating water bath and monitored with a
thermocouple placed as close as feasible to the point where the IR
radiation was focused.

FT-IR INSTRUMENTATION AND DATA ACQUISITION. Most spectra were
recorded with a Mattson Instruments Sirius 100 FT-IR spectrometer

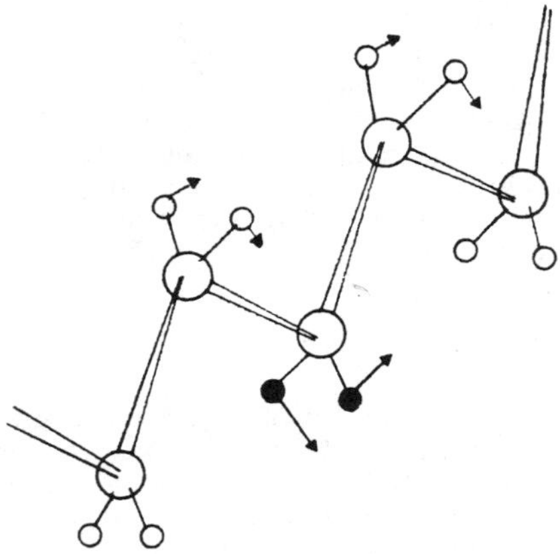

FORM OF THE CD_2 ROCKING MODE
IN A TRANS-POLYETHYLENE SEGMENT

Figure 1. Form of the CD_2 Rocking Mode. The deuterium atoms on the central CD_2 group are depicted as filled-in circles. Arrows are displacement vectors. (Reprinted from ref. 19. Copyright 1973 American Chemical Society.)

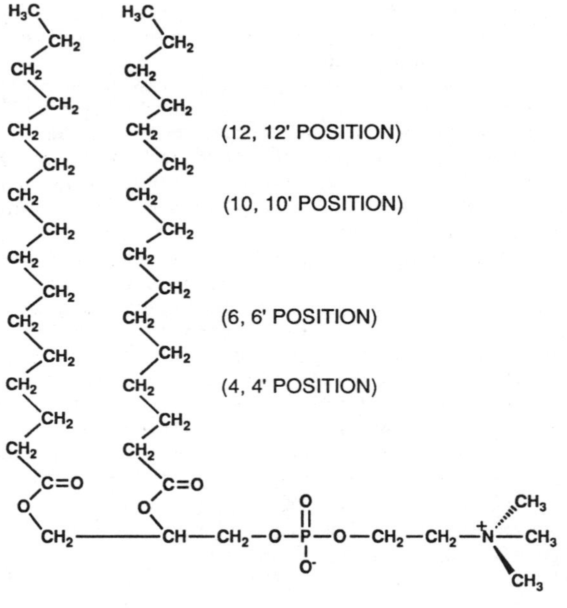

PRIMARY STRUCTURE OF 1, 2 DIPALMITOYLPHOSPHATIDYLCHOLINE
SPECIFICALLY DEUTERATED DERIVATIVES WERE PREPARED AT THE
(4, 4'), (6, 6'), (10, 10'), AND (12, 12') POSITIONS AS SHOWN.

Figure 2 : Structure of Dipalmitoylphosphatidylcholine (DPPC)

(Mattson Instruments, Madison,WI) equipped with a HgCdTe detector (1 x 1 mm) specially fabricated (Infrared Associates, Cranbury, NJ) for high response near 620 cm^{-1}. Typically, 2000 scans of sample and background were separately co-added and apodized with a triangular function at a spectrometer resolution of 4 cm^{-1}. Interferograms were zero-filled (2 levels) and Fourier transformed to yield data encoded at ~1 cm^{-1} intervals. For samples containing solely aqueous dispersions of phospholipids, a parallel series of spectra were obtained by Dr. R.A. Dluhy on a Digilab FTS-40 instrument located at the Batelle Laboratories (Columbus, Ohio) National Center for Biomedical FT-IR. These IR spectra were collected by co-addition of 1024 scans at 2 cm^{-1} resolution.

FT-IR DATA REDUCTION. The CD$_2$ rocking modes (610-660 cm^{-1}) in the current application appear as very weak features (0.040-0.0001 Absorbance units) on an intense (0.4-1.0 Absorbance units), temperature-dependent, sloping D$_2$0 libration background. In addition,both cholesterol and Gramicidin D contribute weak spectral features in this area. Thus,careful background subtraction is mandatory. This was initiated with samples of fully proteated DPPC (with matched levels of cholesterol, Gramicidin D and D$_2$0), measured under the same conditions of cell path length, sample D$_2$0 content, and temperature, as a reference. Spectral subtraction is accomplished by fitting a cubic polynomial function (J.W. Brauner, unpublished results) to the sloping background of each sample and background spectrum between 665 and 595 cm^{-1}. The function was selected to produce spectra with minimal residual slope and/or curvature.

For cholesterol-containing samples, background spectra are scaled using a cholesterol feature at 609 cm^{-1}, and subtracted from sample spectra at corresponding temperatures. A final linear baseline correction is then applied, if necessary, using standard instrument software.

Incomplete resolution of gauche CD$_2$ rocking modes along with the occasional presence of incompletely subtracted, weak spectral features arising from cholesterol or Gramicidin D, made curve fitting for quantitative analysis of band intensities necessary. Band profiles were constructed from addition of individual Gaussian-Lorentzian bands using an interactive algorithm (J.Brauner, unpublished results) constructed specially for this purpose. Areas of the component bands were computed and used for calculation of the relative intensities of trans and gauche conformers.

RESULTS

Typical raw (neither subtracted, nor baseline - flattened) data are shown for 6-d$_4$DPPC in Figure 3, while the appearance of the data following the manipulations described in the experimental section are shown in Figure 4. The data shown for a temperature (49°C) well above T$_m$ reveals the tt band as the strongest spectral feature near

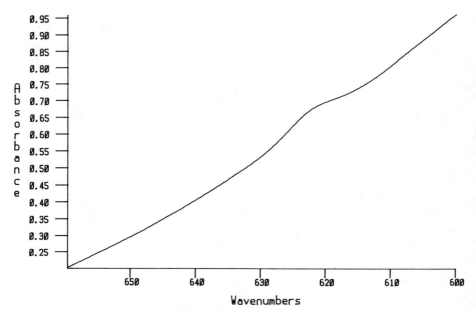

Figure 3 : Spectrum of the CD_2 Rocking Mode Region of DPPC Before
Water Subtraction. The spectrum shown is for gel phase
6-d_4 DPPC.

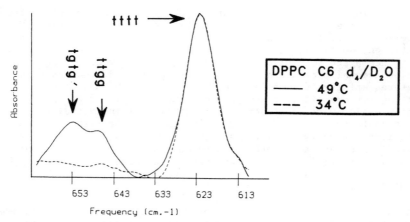

Figure 4 : Spectrum of 6-d_4 DPPC Below (34°C) and Above (49°C) the
Main Phase Transition.

622 cm^{-1}, along with the g̲t̲ gauche doublet at 646 and 652 cm^{-1}. The spectra are indicative of substantial conformational disorder in the acyl chains. Cooling of the system below 41° produces a highly ordered gel phase, where the evident weakness of the gauche doublet at 34°C , as shown in Figure 4, reveals the highly ordered nature of the gel phase.The quantitative results for disorder at various chain positions and temperatures are summarized in Table III.

An interesting phenomenon is noted at the 4 (4') positions of both DPPC and dipalmitoylphosphatidylethanolamine (DPPE) in the gel phases. Substantial conformational disorder is observed. The data for the 4-d$_4$ DPPE are shown in Figure 5. The observation of 6.7% disorder is consistent with the data for 4-d$_4$ DPPC (Table III) below T$_m$. Possible origins of this somewhat unexpected result are suggested below.

Table III : Comparison of Percent of Total Gauche Conformers for Pure DPPC, 2:1 DPPC/Cholesterol, and DPPC/Gramicidin Mixtures.

			Temperature		
derivative	5	20	34	44	49
4-d$_4$[a]		7.6	10.9±6.1		20.7±4.2
4-d$_4$/cholesterol[b]			1.5±1.0		3.7±1.0
6-d$_4$[a]			1.7±0.3		32.3±2.3
6-d$_4$/cholesterol[b]			2.6±1.0		3.6±1.0
6-d$_4$/Gramicidin(30:1)[c]	1.0±0.5		2.6±1.3	17.0±3.4	
6-d$_4$/Gramicidin(10:1)[c]	5.8±3.0		9.8±3.3	15.0±3.0	
10-d$_4$[a]			2.5±1.0		19.7±0.8
12-d$_4$/cholesterol[b]	4.5±1.0	7.5±1.0	6.7±2.0	10.1±2.5	10.6±2.0

[a]Data from reference 23.
[b]Data from reference 24.
[c]Data from Reference 25.

The effects of cholesterol at various bilayer depths are shown in Figures 6 and 7. Figure 6 compares the CD$_2$ rocking modes for 6 and 12-d$_4$ DPPC at 34°C, while the same molecules are compared at 49°C in Figure 7. The results of the quantitative analysis are included in Table III. Most notable in Figure 4 and Table III are the suppression of gauche rotamer formation at 49°C at the 6(6') position, as shown by the relative weakness of the g̲t̲ features at 646 and 652 cm^{-1} in the cholesterol containing samples. The strong position dependence of cholesterol ordering is also directly illustrated in Figure 7. At 49°C 12-d$_4$ DPPC has 10.6 % gauche rotamers and 6-d_4 DPPC has 3.6%. For comparison, we note that the cholesterol-free 6-d_4 derivative has about 32 % disorder at the 6(6') position (Table III).

Gramicidin D is a naturally occurring mixture of three closely related derivatives, including 87 % Gramicidin A, a pentadecapeptide antibiotic of primary sequence:
HCO-(L)-Val-Gly-(L)-Ala-(D)-Leu-(L)-Ala-(D)-Val-(L)-Val-(D)-Val-((L)-Trp-(D)-Leu-)$_3$-(L)-Trp-NHCH$_2$-CH$_2$OH . The molecule has been widely investigated as a model for lipid/protein interaction with membrane proteins (28) and is known to span the membrane bilayer as a β-helical pore dimer (29). Spectra of 6-d_4-DPPC /Gramicidin D complexes at mole ratios of 30/1 and 10:1 are shown at several temperatures in figures 8 and 9 respectively. The quantitative effects of Gramicidin D at the 6(6') position of DPPC are summarized in Table III. At high protein levels in the gel phase, the percentage of gauche rotamers is substantially enhanced from the protein-free bilayers. In the liquid crystal phase, the disordering is somewhat reduced from pure 6-d_4 DPPC.

DISCUSSION

Pure Phospholipid Bilayers. The assumptions underlying the current quantitative analyses of conformational disorder are as follows:

(i) The relative absorptivities of the 622, 646 and 652 cm^{-1} marker bands are unchanged in going from alkane to phospholipid environments.
(ii) The relative extinction coefficients are also unaltered by the presence of cholesterol or Gramicidin D. As these vibrational modes (vide supra) are relatively highly localized, assumptions (i) and (ii) seem appropriate.
(iii) The molecular motions to which the CD$_2$ rocking modes are primarily sensitive are trans-gauche isomerization. The rapid time scale of molecular vibrations "uncouples" these frequencies from the slower motions seen by NMR and other spectroscopic methods. While the linewidth of the individual vibrational modes will respond to more collective slower motions of the acyl chains, the relatively small temperature regime explored in the current work ensures that the effect will be small.

The questions which this study addresses are enumerated below:
(i) What fractions of trans and gauche conformers are present in the acyl chains for various bilayer physical states ?
(ii) What is the depth dependence of the trans/gauche population ratio along the acyl chain and how is it affected by cholesterol, a molecule suspected of having differential effects at different depths in the bilayer ?
(iii) What is the quantitative effect of Gramicidin D, a widely used model for intrinsic membrane proteins, on acyl chain conformational order ?
(iv) What is the nature of the conformational disorder (kinks, single gauche rotamers, multiple gauche forms, etc.) in all the circumstances alluded to above ?

The current results for pure phospholipid bilayers (Table IV) may be compared with other physical estimates of acyl chain

order. This requires multipication of the average of the
experimental numbers at the 4,6 and 10 positions (Table III) by the
appropriate number of C-C bonds (excluding the $C_{15}-C_{16}$ bond). This
results in 3.6 gauche bonds/chain at 49°C, a value which is a lower
limit since more conformational flexibility is anticipated towards
the bilayer center. The maximum allowed value for fractional
disorder ($g^+ + g^-$) for a C_{16} random coil alkane at 49°C is 37.1% (as
calculated from the rotational isomeric state model (22)). Assuming
this value for bonds 11-15 yields an upper limit of 4.2 gauche
bonds/chain. Our range of 3.6-4.2 is compared with those determined
from other physical methods (Table IV). The current estimate is in
excellent accord with the dilatometric data of Nagle and Wilkinson
(30). The agreement with the NMR data of Petersen and Chan(31) and
the Seeligs (7) is fair, but the NMR determination of conformational
disorder is very dependent on the dynamic model selected. A
substantial advantage of the current FT-IR determination lies in its
ability to monitor the depth dependence of the disorder.

Table IV: Conformational order in Liquid Crystalline DPPC by Various
Techniques.

Technique	Result	Reference
FT-IR	3.6-4.2 gauche/chain	current work
^{2}H NMR	3-6 gauche/chain	7
Dilatometry	3.8 gauche/chain	30
^{2}H NMR	P_{trans} (upper part of chain)	31
	= 0.8-0.9	
Raman	about 5/chain above T_m	16

DPPC/Cholesterol Mixtures. To explore this advantage, an obvious
starting point is the investigation of cholesterol/DPPC systems.
Many spectroscopic investigations (for a comprehensive,recent review
see 32), have led to general acceptance of the fact that
cholesterol, upon insertion into phospholipids, fluidizes and/or
disorders PC's below T_m, while rigidifying phospholipids above T_m. A
simple, specific, structural model has been proposed (33) in which
phospholipid conformational disorder is restricted by the sterol
4-ring nucleus for a distance of 10-11 Angstroms toward the bilayer
center from the interfacial region. In contrast, those acyl chain
segments that protrude beyond this bulky region of sterol toward the
bilayer center, are thought to possess enhanced conformational
flexibility due to poor packing between themselves and the sterol
side chain. The quantitative aspects of this model, as determined in
the current experiments, are heretofore unexplored. The ability of
cholesterol to strongly restrict gauche rotamer formation at the 4
and 6 acyl chain positions above T_m (Table III) is remarkable. At

the former position at 49°C, the % of total gauche conformers is
20.7 +/- 4.2 and 3.7 +/- 1.0 in the absence and presence (2:1 mole
ratio) of cholesterol respectively, i.e. a reduction of a factor of
6. Similarly a factor of 9 reduction is noted at the six position.
In contrast ,conformational disorder is much less restricted at the
12 position of DPPC where 10.6 +/- 2.0 % gauche conformers are
observed.
    The disordering effects of cholesterol on the lipid gel phase
are much more difficult to measure because of the weakness of the
gauche marker bands. Discussion of the 4 position will be deferred
because of the anomalously high disorder observed in the gel phase
of the cholesterol free system. The 6-d$_4$ DPPC derivative showed
(Figure 6) values of 1.7 +/- 0.3 and 2.6+/- 1.0 at 34°C in the
absence and presence of cholesterol. These numbers are the same
within the current experimental precision. It appears that a real
increase occurs further down the chain where the 10 position
(cholesterol-free) has 2.5 +/-1.0 disorder and the 12 position
(cholesterol-containing) has 6.7 +/- 2.0. Further experimental work
and completion of control experiments (which await completed
syntheses) are needed. The general features of the proposed model
for cholesterol insertion are consistent with,and extended by,the
quantitative picture produced in the current study.
    The experimental data contain two puzzling features. First, the
observation of substantial disorder 4(4') position in both DPPC and
DPPE (Table III), was unexpected. There is NMR precedence for the
result for DPPE. Blume et al (34) have modelled the gel phase ^{2}H NMR
spectrum of 4-d$_4$ DPPE with a six site jump model with equal
probabilities for three sites of the rotational jump and variable
probabilities for the trans and gauche sites. Gauche probabilities
for the 4 position ranged from 5 % at 25°C to 7% at 55°C, in good
accord with the FT-IR observation reported here. Similar
considerations may apply to DPPC although is noted that the NMR
simulations predict greater disorder toward the center of the
bilayer in the gel phase of DPPE. Our experimental results for DPPC,
in contrast, show the greatest disorder at the 6 position in the
bilayer liquid crystal phase.
    The second puzzling feature is the observation of more liquid
crystal phase disorder at the 6 position than at either the 4 or 10
positions of DPPC (Table I). The results appear statistically
significant and contrast with the observed (35) NMR order parameter
profile (constant from positions 2→10, diminished order from
position 10 to the bilayer center). Further experiments with DPPC
derivatives deuterated at positions closer to the bilayer center may
clarify the situation.

DPPC/Gramicidin D Mixtures.   Experiments to determine the effects
of Gramicidin D insertion on lipid conformational order, also
illustrate one of the important advantages of the CD$_2$ probe method;
namely, the ability to discern between various sources of disorder
(Table V). Spectra of DPPC/Gramicidin mixtures at various

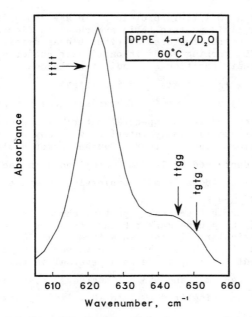

Figure 5 : Gel Phase CD$_2$ Rocking Mode Spectrum of 4-d$_4$ DPPE. The molecule has its main phase transition at 64°C. The residual gel phase disorder is substantial.

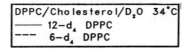

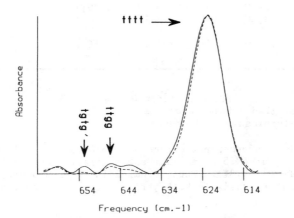

Figure 6 : CD$_2$ Rocking Mode Spectrum of 6-d$_4$ DPPC and 12-d$_4$ DPPC in the Presence of Cholesterol at 34°C

temperatures for a 30:1 lipid/protein ratio are shown in Figure 8; those for a 10:1 lipid protein ratio are shown in Figure 9. The expected three spectral features attributable,as detailed previously, to CD_2 rocking modes in particular conformational states at the C6 acyl chain position are indicated. In addition to bands at 622 (t<u>ttt</u>), 653 (g'<u>tgt</u> + t<u>tgt</u>), and 649 (t<u>tgg</u>) cm^{-1}, a weak feature is noted (Figure 7) at 643 cm^{-1} which arises from a Gramicidin band not completely removed by the spectral subtraction process; the band however, does not substantially interfere with any of the trans-gauche conformational markers. Gauche marker positions in the 30:1 sample are decreased by 1-2 cm^{-1} from those in the 10:1 sample. Finally, CHD rocking modes from incompletely deuterated lipid appear at 657 cm^{-1} in Figure 8.

The spectra exhibit several important characteristics. First, the intensity of the gauche marker bands increases with increasing temperature, particularly in the range from 34-44 $^{\circ}$C. Second, the overall fraction of gauche conformers increases with increasing Gramicidin content. Third, there is a reversal in the relative intensity of the 649 cm^{-1} and 653 cm^{-1} markers (most easily observed at 34 and 44 $^{\circ}$C) on going from 30:1 to 10:1 lipid/protein ratio (Table V).

Table V : Percent of Gauche Conformers in tgt and tgg$^{\underline{d}}$ Classes

Mixture	tgt Class (%)		tgg Class (%)	
	34°C	44°C	34°C	44°C
6-d$_4$DPPC	37±10	74±18^a	62±10	25±18
10:1^b	20±10	15±10	79±10	84±10
30:1^b	53±10	78±16	46±10	21±16
12-d$_4$/chol^c	33±10	54±15	66±10	45±15

^{a}49°C
^{b}mole ratios of 6-d$_4$ DPPC/Gramicidin D
^{c}mole ratio of 12-d$_4$/cholesterol = 2:1
^{d}tgg class determined by difference

The gauche conformer percentage at the 6 position in the 30:1 mixture is 17.7% at 48.5°C, a value substantially reduced from peptide-free DPPC (32.3%). Thus, Gramicidin clearly orders the liquid crystalline phase. In both pure 6-d$_4$ DPPC in the liquid

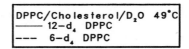

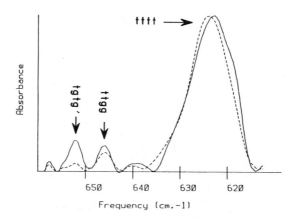

Figure 7 : CD₂ Rocking Mode Spectrum of 6-d₄ DPPC and 12-d₄ DPPC in the Presence of Cholesterol at 49°C

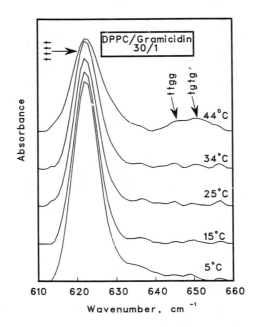

Figure 8 : CD₂ Rocking Mode Spectrum of 6-d₄ DPPC in the Presence of Gramicidin (30:1 lipid:peptide, mol:mol)

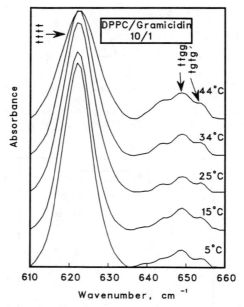

Figure 9 : CD$_2$ Rocking Mode Spectrum of 6-d$_4$ DPPC in the Presence
of Gramicidin (10:1 lipid:peptide, mol:mol)

crystal phase and at low levels of Gramicidin D (30:1, mol:mol), the disorder is due primarily to kinks and/or single gauche rotamers ($78\pm16\%$ of the total gauche conformer population), as judged by the greater intensity of the 652 cm^{-1} feature. Jackson (37) and Marsh (38) have suggested that kinks are most likely to be favored,as they induce less disruption in phospholipid intermolecular packing.

The situation is dramatically altered at high protein levels. For 10:1 DPPC/Gramicidin D, multiple gauche forms dominate at all temperatures. For example, kinks and single gauche bends make up only $15\pm10\%$ of all gauche conformers at $44^{o}C$. It is noted that increasing the temperature from 34 to $44^{o}C$ does not alter the relative intensities of the gauche marker bands in the 10:1 mixture. In contrast, the 652 cm^{-1}/622 cm^{-1} ratio increases much more rapidly than the 646 cm^{-1}/622 cm^{-1} ratio in the 30:1 mixture. Thus, high levels of protein create large defects in the bilayer structure, even at low temperature, which can be filled most efficiently by multiple gauche (646 cm^{-1}) disordered forms. Similar behavior is noted for DPPC/cholesterol (2:1, mol:mol) complexes. The temperature dependence of the 652/646 ratio has been studied at both the 12 and 6 positions for CD_2 groups of the acyl chains. At low temperatures in both cases, the multiple gauche forms dominate (Table V).

However, as the temperature increases, the 652 cm^{-1} marker increases in intensity at a much faster rate. This suggests the presence of packing defects filled by multiple gauche forms even at low temperatures. Further temperature -induced disordering occurs, as in the cholesterol-free system, by formation of either kinks or single gauche rotamers, with the former being the more likely source of conformational alteration. These data portray a detailed picture of the extent and nature of conformational disordering.

Advantages of the CD_2 Probe Method.   The advantages of the CD_2 infrared probe method utilised here for studies of conformational disorder are:
(i)   The probe molecules needed (CD_2 vs CH_2) do not effect the physical properties of the bilayers into which they are inserted.
(ii)   The method senses ,to a large degree, only acyl chain conformational disordering, as the vibrational modes are essentially insensitive to slower motions, especially over the narrow temperature interval studied.
(iii)   Most importantly, the depth dependence of disordering in the bilayer is directly monitored.
(iv)   Data analysis, under the assumptions listed previously concerning transferability of the extinction coefficients from alkanes to bilayer environments, is straightforward and independent of any mathematical model for acyl chain motions.
The drawbacks of the method are first, the requirement for

specifically deuterated phospholipids and second, the weakness of
the CD_2 rocking modes. Current synthetic approaches, although
tedious, permit 2-3 grams of material to be made in a relatively
straightforward fashion. Second, the weakness of the bands is
unlikely to be readily overcome. Increasing cell path lengths raises
background absorbance to an unacceptable level. Reflectance methods
for data acquisition failed in numerous attempts. Tunable diode
laser spectroscopy may afford certain advantages,although the
technology is still (as of 7/90) not perfected in this spectral
region.

ACKNOWLEDGMENTS

This work was supported by the Public Health Service through
grant GM-29864 to RM. Further support for the update of FT-IR
instrumentation came via the Busch bequest to Rutgers University. We
thank Prof. A. Blume for a sample of $4-d_4$DPPE.

LITERATURE CITED

1. Singer, S.J.; Nicolson, G.L. Science 1972, 175, 720-31.
2. McElhaney, R.N. In Membrane Fluidity; M.Kates; L.A.Manson, Eds.;
Plenum : New York, 1984, pp. 249-278.
3. Carruthers, A.; Melchior, D.L. Trends in Biochemical Sciences
1986, 11, 331-5.
4. Taylor, M.O.; Smith, I.C.P. Biochim. Biophys. Acta 1980, 59,
140-9.
5. Brasitus, T.A.; Tall, A.R.; Schachter, D. Biochemistry 1980, 19,
1256-61.
6. Yamaguchi, T.; Koga, M.; Fujita, Y.; Kimoto, E. J. Biochem. 1982
91, 1299-1304.
7. Seelig, A.; Seelig, J. Biochemistry 1974, 13, 4839-45.
8. Seelig, J.; MacDonald, P.M. Acc. Chem. Res. 1987, 20, 221-8.
9. Bloom, M.; Smith, I.C.P. In Progress in Protein-Lipid
Interactions; A.Watts; J.J.H.H.M De Pont, Eds.; Elsevier :
Amsterdam, 1985; Vol. 1, pp. 61-89.
10. Westerman, P.W.; Vaz, M.J.; Strenk, L.M.;.Doane,J.W. Proc. Nat.
Acad. Sci. U.S.A. 1982, 79, 2890-94.
11. Wittebort, R.J.; Olejniczak, E.T.; Griffin, R.G. J.Chem.Phys.
1987, 86, 5411-20.
12. Mendelsohn, R.; Mantsch, H.H. In Progress in Protein-Lipid
Interactions; Watts, A.; DePont, J.J.H.H.M.. Eds.; Elsevier :
Amsterdam, 1986; Vol. 2, pp.
13. Dluhy, R.A.; Moffatt, D.; Cameron, D.G.; Mendelsohn, R.;
Mantsch, H.H. Can.J. Chem. 1985, 63, 1925-32.
14. Dluhy, R.A.; Mendelsohn, R.; Casal, H.L.; Mantsch, H.H.
Biochemistry 1983, 22, 1170-7.   '
15. Gaber, B.P.; Peticolas, W.L. Biochim. Biophys. Acta 1977, 465,
260-74.
16. Pink, D.A.; Green, T.J.; Chapman, D. Biochemistry 1980, 19,
349-56.
17. Vogel, H.; Jahnig, F. Chem.Phys. Lipids 1981, 29, 83-101.
18. Brown, K.G.; Peticolas, W.L.; Brown, E. Biochem. Biophys. Res.
Comm. 1973, 54, 358-67.
19. Snyder, R.G.; Poore, M.W. Macromolecules 1973, 6, 708-15.

20. Maroncelli, M.; Strauss, H.L.; Snyder, R.G. J.Chem.Phys. 1985, 82, 2811-24.
21. Maroncelli, M.; Strauss, H.L.; Snyder, R.G. J.Phys.Chem. 1985, 89, 4390-5.
22. Flory, P.J. Statistical Mechanics of Chain Molecules; Hanser : Munich, 1989.
23. Mendelsohn, R.; Davies, M.A.; Brauner, J.W.; Schuster, H.F.; Dluhy, R.A. Biochemistry 1989, 28, 8934-9.
24. Davies, M.A.; Schuster, H.F.; Brauner, J.W.; Mendelsohn, R. Biochemistry 1990, 29, 4368-73.
25. Davies, M.A.; Brauner, J.W.; Schuster, H.F.; Mendelsohn, R. Biochem. Biophys. Res. Comm. 1990, 168, 85-90.
26. Tulloch, A.P. Chem.Phys. Lipids 1979, 24, 391-406.
27. Kodama, M.; Kuwabara, M.; Seki, S. Biochim. Biophys. Acta 1982, 689, 567-70.
28. de Kruijff, B.; Cullis, P.R.; Verkleij, A.J.; Hope, M.J.; van Echteld, C.J.A.; Taraschi, T.F.; van Hoogevest,P.; Killian, J.A.; Rietveld, A.; van der Steen, A.T.M. In Progress in Protein-Lipid Interactions; Watts, A.; DePont, J.J.H.H.M., Eds.; Elsevier : Amsterdam, 1985
29. Urry, D.W.; Trapane, T.L.; Prasad, W.U. Science 1983, 221, 1064-7.
30. Nagle, J.F.; Wilkinson, D.A. Biophys.J. 1978, 23, 159-75.
31. Petersen, N.O.; Chan, S.I. Biochemistry 1977, 16, 2657-67.
32. Presti, F.T. In Membrane Fluidity in Biology, Cellular Aspects; Aloia, R.C.; Boggs, J.M., Eds.; Academic Press : Orlando, FL., 1985, Vol. 4.
33. Rothman, J.E.; Engleman, D.M. Nature New Biology (London) 1972, 237, 42-4.
34. Blume, A.; Rice, D.M.; Wittebort, R.J.; Griffin, R.G. Biochemistry 1982, 24, 6220-9.
35. Davis, J.H. Biophys.J. 1979, 27, 339-58.
36. Shannon, V.L.; Strauss, H.L.; Snyder, R.G.; Elliger, C.A.; Mattice, W.L. J.Am. Chem. Soc., 111, 1947-58.
37. Jackson, M.B. Biochemistry 1976, 15, 2555-61.
38. Marsh, D. J. Membrane Biol. 1974, 18, 145-62.

RECEIVED August 2, 1990

Chapter 3

Pressure Tuning Fourier Transform Infrared and Raman Spectroscopy of Anionic Surfactants

Patrick T. T. Wong and Henry H. Mantsch

Steacie Institute of Molecular Sciences, National Research Council of Canada, Ottawa, Ontario, K1A 0R6, Canada

In this report we describe novel pressure tuning vibrational spectroscopic techniques that can be used to study aqueous surfactant solutions and discuss in some detail two examples of such studies with micellar solutions of anionic surfactants, one using Fourier transform infrared (FT-IR) and another using Raman spectroscopy.

Most of the experimental and theoretical work on the aggregation of ionic surfactants in water has been devoted to understanding how this phenomenon is affected by such factors as concentration, temperature or chemical nature of the surfactants. Much less is known as to how surfactant aggregation is affected by an increase in hydrostatic pressure. Advances in the technique of high pressure vibrational spectroscopy (FT-IR and Raman) of aqueous systems have allowed us now to examine the effect of hydrostatic pressure on the structural and dynamic properties of a large number of surfactants in solution.

There are fundamental reasons why to study the effect of pressure on surfactant molecules in solution. While an increase in hydrostatic pressure is similar to a decrease in temperature, it is often not fully appreciated that through a change in pressure one solely changes the space available to the molecules under investigation, whereas through a change in temperature one varies in fact two parameters: the energy of the molecules and the space available to them. Therefore, different surfactant molecules will respond to external pressure differently when the intermolecular interactions are varied. This also applies to the same anionic surfactants with different counter cations. Consequently, with the addition of a pressure parameter, new information can be obtained on the structure and dynamics of the surfactant molecules in aqueous solution.

0097–6156/91/0447–0044$06.00/0

Experimental Methods

To obtain vibrational spectra tuned by external pressure one needs - in addition to the infrared or Raman spectrometer - a special high-pressure optical sample cell, an optical interface between the sample cell and the spectrometer, and a device to measure the pressure on the sample [1].

There are two basic types of high pressure optical cells, namely the window cell and the opposed anvil cell. The window cell is essentially a metallic high pressure vessel with optical windows. Materials such as diamond, sapphire, glass, quartz, silicon or Irtran can be used as optical windows for high pressure studies. The selection of a particular type of window material is based on its mechanical strength - which will limit the maximum pressure that can be achieved, and by the spectroscopic range over which spectra may be collected. Here at the NRC in Ottawa, we have designed and built a window cell for Raman spectroscopic studies with three glass windows [2]. With a specially designed sample holder [3] this cell is suitable for recording spectra of aqueous solutions. The maximum pressure that can be generated in this cell is 25 kbar, which is the highest pressure obtained so far with a window cell. Moreover, to our knowledge this is the only cell that can be used to obtain polarized Raman spectra under high pressure [3,4].

The opposed anvil cell consists of two optical anvils and a gasket, located between the parallel faces of the two opposing anvils. Samples are placed in the hole of the gasket and are pressurized when the opposed anvils are pushed towards each other. The most common material for anvils is diamond. For mid and far infrared spectra, type IIa diamonds are used, while low-fluorescent type Ia diamonds are used for Raman spectroscopic measurements [5]. We have also devised a glass anvil cell for Raman spectroscopic measurements [6], and a calcium fluoride anvil cell for infrared spectroscopic measurements [7] with attainable working pressures of 13 and 6 kbar, respectively. Diagrams, for the interested reader, of the window and opposed anvil cells can be found in reference 1.

In view of the small sample size, the optical aperture used for pressure tuning vibrational spectroscopic measurements is also very small. This does not create serious problems for high pressure Raman spectroscopic measurements since the laser beam can be focused to 30-40 μm. Thus, the optical system employed in a standard Raman scattering experiment can be used and no special optical interface is required for the pressure tuning Raman spectroscopic measurements.

The situation, however, is different for the infrared spectroscopic measurements with opposed anvil cells. The source beam in commercial Fourier transform infrared spectrometers is generally focused to about 1 cm diameter at the sample, whereas the diameter of the gasket hole in the high pressure cell is only about 0.3 mm. Therefore, a source beam condensing system is required in order to obtain infrared spectra with a good signal-to noise ratio. Commercial beam condensers (4X, 6X) could, in principle, be adapted for these purposes. In practice, however, the mirrors of the commercial beam condensors are difficult to align and the working space between the mirrors is too small for many anvil cells. We have therefore developed a special beam condensing system which uses sodium chloride optical lenses [8].

Several methods exist for measuring pressures in optical cells, of which many have been developed in this laboratory. For monitoring the pressure in a window cell, we generally use a gauge consisting of a high precision manganin resistance coil [2]. The pressure is determined from the change in resistance of the manganin coil which was originally calibrated against a controlled clearance 20 kbar pressure balance with a tungsten carbide piston and cylinder.

For monitoring the pressure in anvil cells we use the frequncy shift of internal, chemically inert pressure calibrants. For Raman spectroscopic measurements, the most commonly used method is based on the pressure-induced frequency shift of the R_1 fluorescence line of a small piece of ruby that is placed in the sample compartment of the cell, next to the sample [1]. For infrared spectroscopic measurements, we have developed a quartz pressure scale [9], a $BaSO_4$ pressure scale [10], and an HOD pressure scale [11]. In the case of the first two techniques, a small amount of powdered quartz or $BaSO_4$ powder are placed in the sample hole on the gasket, together with the sample under investigation. The infrared spectra of quartz or $BaSO_4$, which are relatively simple, are recorded simultaneously with the spectrum of the sample and the pressure on the sample is then determined from the frequency shift of the infrared bands of quartz or $BaSO_4$. The HOD pressure scale was developed specifically for aqueous solutions. In this case, the pressure in solution is determined from the frequency shift of the uncoupled O-H stretching band of residual HOD in D_2O solutions, or from the uncoupled O-D stretching band of residual HOD in H_2O solutions [11].

Examples

Two representative studies are described here which reveal the type of structural information that can be gained from pressure tuning infrared and Raman spectroscopic experiments with micellar solutions of anionic surfactants.

Study # 1. Micelles of Sodium and Potassium Decanoate. Infrared spectra of micellar solutions of sodium decanoate (NaC_{10}) and potassium decanoate (KC_{10}) in D_2O (15 wt%) were obtained as a function of increasing hydrostatic pressure up to pressures exceeding 50 kbar. These measurements were performed with a conventional Fourier transform infrared spectrometer (Bomem, Model DA 3.02) and a home build diamond anvil cell, using quartz as the internal pressure calibrant. Infrared spectra of NaC_{10} and KC_{10} at several pressures are shown as stacked contour plots in Figures 1, 2 and 3. The spectral regions shown here contain various vibrational modes of diagnostic value: the region of the C-H stretching modes, the region of the methylene bending modes and the region of the carboxylate stretching modes.

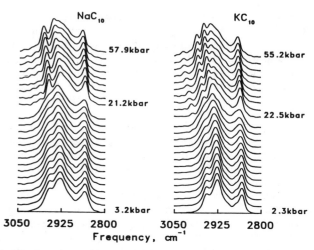

Figure 1. Stacked contour plots of infrared spectra of NaC_{10} and KC_{10} in the region of the CH stretching bands as a function of external pressure. (Reproduced with permission from ref. 15. Copyright 1989 Academic Press.)

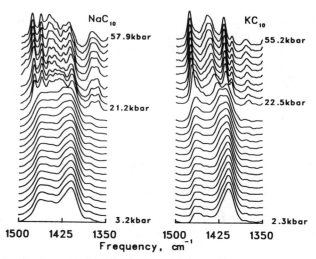

Figure 2. Stacked contour plots of infrared spectra of NaC_{10} and KC_{10} in the region of the CH_2 bending bands as a function of external pressure. (Reproduced with permission from ref. 15. Copyright 1989 Academic Press.)

Dramatic changes are evident at 21.2 kbar in the infrared spectrum of NaC_{10} and at 22.5 kbar in the spectrum of KC_{10}. The frequencies of most infrared bands of the two surfactants undergo a discontinuous shift at these critical pressures. This is illustrated in Figure 4 for the C-H stretching bands and in Figure 5 for the CH_2 bending bands. These frequencies were obtained from infrared spectra subjected to band narrowing by Fourier derivation [12] and are accurate to better than 0.1 cm^{-1} [13]. The spectral changes in Figures 4 and 5 identify the barotropic event at 21.2 kbar for NaC_{10} and at 22.5 kbar for KC_{10} as a pressure-induced phase transition from the isotropic micellar phase to a solid-like coagel phase [1,14]. We refer to these pressures as the critical coagelization pressures [15]. It is clear that the critical coagelization pressure of the decanoate micelles is significantly affected by the nature of the counterion, the difference between NaC_{10} and KC_{10} being 1.3 kbar. This is the result of the different magnitude of interaction between the sodium or potassium cation and the carboxylate group of the surfactant in the micellar and in the coagel phases of NaC_{10} and KC_{10}.

At atmospheric and low pressure, all the infrared bands of NaC_{10} and KC_{10} are broad, reflecting the conformationally and orientationally disordered character of the surfactants in the micellar phase [14-16]. The broad infrared bands are maintained in the micellar phase at pressures up to the critical coagelization pressure. This demonstrates that the surfactant molecules are kept in a conformationally and orientationally disordered form and that external pressure does not induce a structural ordering of the micellar phase. On the contrary, an increase in external pressure induces more conformational disorder of the methylene chains in NaC_{10} and KC_{10} micelles. This is evident from a continuous broadening of the methylene chain bands in the micellar phase as the pressure increases from atmospheric pressure to the critical coagelization pressure (see methylene band contours in Figures 1 and 2). On the other hand, the conformational disorder of the carboxylate groups in the micellar phase is not significantly affected by external pressure. This is demonstrated by the fact that the width of the carboxylate bands remains practically constant in the micellar phase up to the critical coagelization pressure (see Figure 3).

Clearly, the infrared spectra of the sodium and potassium decanoate micellar solutions are considerably different, as are their pressure dependencies. Since the only difference between these two micellar systems is the size, and thus the charge density of the counterions, the different infrared spectra must be taken as evidence that in the alkali decanoate micelles the sodium or potassium counter cations interact differently with the carboxylate groups of the surfactant molecules.

Infrared bands that can provide information on ion-surfactant interactions are the two carboxylate bands, the antisymmetric COO^- stretching band between 1545 and 1565 cm^{-1} which is mainly due to the $C=O$ stretching mode ($\nu C=O$), and the symmetric COO^- stretching band around 1410 cm^{-1} which is mainly due to the CO^- single bond stretching (νCO^-). The methylene bending bands (δCH_2) between 1465 and 1480 cm^{-1} are diagnostic of the chain packing pattern.

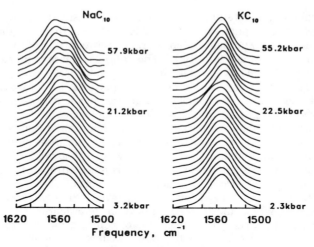

Figure 3.   Stacked contour plots of infrared spectra of NaC_{10} and KC_{10} in the region of the C=O stretching bands as a function of external pressure. (Reproduced with permission from ref. *15*. Copyright 1989 Academic Press).

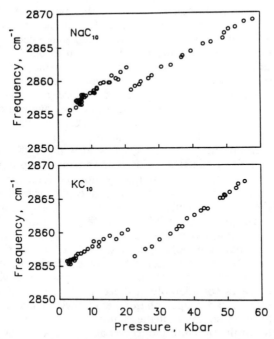

Figure 4.   Pressure dependence of the frequencies of the symmetric CH_2 stretching bands of NaC_{10} and KC_{10}. (Reproduced with permission from ref. *15*. Copyright 1989 Academic Press).

At ambient temperature and pressure the intensity of the νCO^- band is much higher in the potassium decanoate micelles than in the sodium decanoate micelles. The peak intensity ratio between the νCO^- and δCH_2 bands is 1.7 in NaC_{10} and 3.5 in KC_{10}. This ratio decreases more rapidly with increasing pressure in KC_{10} than in NaC_{10} micelles. Just below the corresponding critical coagelization pressure it becomes 1.3 for NaC_{10} and 1.4 for KC_{10}. Because pressure enhances the interactions between the negatively charged carboxylate groups of the surfactant molecules and the positively charged cations, this leads to a partial neutralization of the negative charge on the carboxylate groups. This, in turn reduces the transition moment of the CO^- stretching vibration and thus the intensity of the νCO^- band. The interaction of the carboxylate group with Na^+ is much stronger than that with K^+ due to the smaller size and larger charge density of the sodium ions. Therefore, the νCO^- band is weaker in NaC_{10} where the counterions interact more strongly. While the carboxylate-cation interactions increase with increasing external pressure, the magnitude of compression is expected to be larger in the KC_{10} micelles with the more weakly interacting counterions. Thus, the pressure-induced reduction in the νCO^- transition moment (and therefore the decrease in the νCO^- band intensity) is larger in the KC_{10} micelles than in the NaC_{10} micelles.

It is evident from Figure 1 that in the micellar state the CH_2 stretching bands of NaC_{10} are broader than the corresponding bands of KC_{10}. This indicates that the conformational structure of the methylene chains in the sodium decanoate micelles is more disordered than that in the potassium decanoate micelles. On the other hand, the pressure-induced broadening of these bands is much more pronounced in KC_{10} than in NaC_{10} and near the critical coagelization pressure the widths of the CH_2 stretching bands of the sodium and potassium decanoate micelles become comparable. These results show that the magnitude of conformational disorder in KC_{10} becomes comparable with that of NaC_{10} near the critical coagelization pressure. Moreover, the carboxylate stretching bands of NaC_{10} are also broader than those of KC_{10} (Figures 2 and 3), which indicates that the orientation of the carboxylate head groups in NaC_{10} micelles is also more disordered when compared to that in KC_{10} micelles. This follows from the difference in the negative charge density on the carboxylate groups between the sodium and potassium decanoate micelles. Again, the reduced charge density of the sodium ions in the NaC_{10} micelles is a result of the stronger counterion interactions in this carboxylate soap. Consequently, the repulsion force among the carboxylate head groups is weaker in NaC_{10} micelles and therefore the reorientational freedom of the molecules - and the flexibility of the methylene chains to form gauche bonds - is greater in the sodium decanoate micelles.

In both the sodium and potassium decanoate soaps the width of most infrared bands decreases considerably at the critical coagelization presssure, which indicates that in the coagel phase the methylene chains are largely extended, and that the orientation of the chains is highly ordered. However, the infrared spectra of the pressure-induced coagel phases of NaC_{10} and KC_{10} are very different (Figures 1, 2 and 3). Therefore, the chain packing and the overall structure of the two carboxylate soaps must differ significantly in the coagel phase. Indeed, in the coagel phase of NaC_{10} the δCH_2 mode splits into two bands, at 1476 and 1464 cm^{-1} (frequencies at 21.2 kbar), whereas in the coagel phase of KC_{10} it is a singlet, at 1471 cm^{-1} (frequncy at 22.5 kbar), and remains a singlet up to 55.2 kbar. The splitting of the δCH_2 band in NaC_{10} is due to correlation field interactions between orientationally nonequivalent methylene chains in the lattice [8]. This shows that in the lamellar coagel phase of sodium decanoate the soap molecules pack in such a way that the zig-zag planes of neighboring methylene chains are nearly perpendicular to each other, but they are aligned parallel to each other in potassium decanoate. Because of the stronger counterion interactions in NaC_{10} the effective charge on the CO$^-$ groups is smaller in NaC_{10} than that in KC_{10} and thus the repulsion force between the carboxylate groups of neighboring molecules is smaller in NaC_{10}. This may be an important factor in allowing the decanoate molecules in NaC_{10} to rotate more freely, which then leads to a perpendicular packing of the methylene chains in the coagel of sodium decanoate, whereas in the coagel of KC_{10} the chains are locked in a parallel packing.

Study # 2. Micelles of Sodium Oleate.   The effect of pressure on a micellar solution of sodium oleate is illustrated here by Raman spectroscopy. Representative Raman spectra of aqueous sodium oleate (13 wt%) in the CH stretching region at various pressures are depicted in Figure 6. These Raman spectra were obtained with a conventional Raman system (Spex Model 1408 double monochromator with 1800 line/mm holographic gratings and a cooled RCA C31034 photomultiplier), excited with the 514.5 nm line from a CRL Model 12 argon ion laser [16]. The sample was contained in a special sample holder designed for aqueous systems [3]. A window cell [2] was used with naphta as the pressure-transmitting fluid. The Raman spectrum at 0.05 kbar is typical of the micellar phase and resembles the Raman spectra of liquid *n*-alkanes or liquid fatty acids [17,18]. At 1.71 kbar, the antisymmetric CH$_2$ stretching mode appears as a sharp band at ~2885 cm^{-1}, while the symmetric CH$_2$ stretching mode, initially observed as a strong band at ~2850 cm^{-1}, becomes weaker and broader. The Raman spectrum at 5.05 kbar is typical of the coagel phase of alkanoate soaps and is identical with the Raman spectrum of sodium oleate at atmospheric pressure at temperatures below the critical coagelization temperature, i.e., 11°C [19]. This demonstrates that a pressure-induced structural phase transition - from the micellar state to the coagel state - takes place at a pressure around 1.7 kbar.

The peak height ratio between the antisymmetric and symmetric CH$_2$ stretching modes H-2880/H-2850 is plotted as a function of pressure in Figure 7. This parameter has been shown to reflect the magnitude of interchain interactions in systems that contain polymethylene chains [20]. An abrupt increase in this ratio, and thus an increase in the interchain interactions, is observed at the pressure-induced transition from the micellar to the coagel phase. Furthermore, it is also evident from Figure 7 that the transition has a considerable pressure hysteresis. Interestingly, pressure [16] and temperature [19] have different effects on the H-2880/H-2850 ratio in the micellar phase of

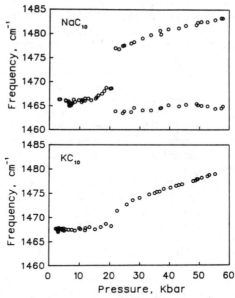

Figure 5.   Pressure dependence of the frequencies of the CH_2 bending bands of NaC_{10} and KC_{10}.

(Reproduced with permission from ref. 15. Copyright 1989 Academic Press.)

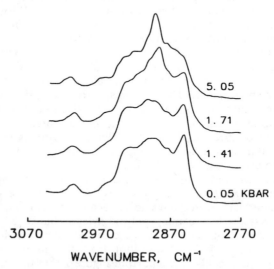

Figure 6.   Raman spectra of aqueous sodium oleate (13 wt%) in the region of the alkyl chain CH stretching vibrations at the indicated pressures. (Adapted from ref. *19*).

sodium oleate. The ratio remains approximately constant when the temperature decreases from 32 to 13°C (which is close to the critical coagelization temperature), whereas the ratio increases by about 17% when the pressure increases from atmospheric pressure to 1.6 kbar (which is close to the critical coagelization pressure). This indicates that an increase in pressure, unlike a decrease in temperature, enhances the interchain interactions in the micellar phase, and denotes that the volume of individual micelles near the transition pressure is smaller than that at atmospheric pressure.

The data acquired from the pressure tuning experiment of sodium oleate micelles, together with data obtained from a similar temperature tuning experiment [16], allow the construction of a temperature versus pressure phase diagram for the aqueous sodium oleate system. This is shown in Figure 8. The solid line represents the critical points of the transition from the micellar phase (M) to the coagel phase (C), while the dashed line represents the critical points of the transition from the coagel phase to the micellar phase. The dotted lines indicate extrapolations of these lines. It is apparent from this phase diagram that the hysteresis for the pressure-induced phase transition is reduced as the temperature increases, while the hystersis for the temperature-induced phase transition is reduced as the pressure increases.

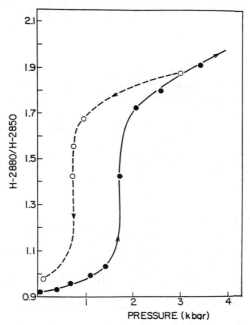

Figure 7.   Effect of pressure on the Raman intensity ratio of the asymmetric and symmetric CH_2 stretching bands of the alkyl chains of sodium oleate in aqueous solutions. Closed symbols indicate measurements with increasing pressure, open symbols show measurements with decreasing pressure. (Adapted from ref. *19*).

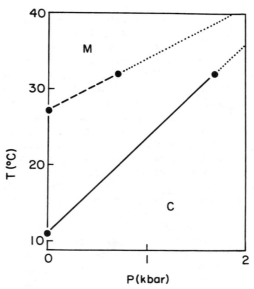

Figure 8.   Temperature versus pressure phase diagram of a micellar solution of sodium oleate, 13 wt % in water. (Adapted from ref. *19*).

Concluding Remarks

Although we have presented here only two examples of pressure tuning vibrational spectroscopic studies with aqueous surfactants, we hope that they are sufficient to demonstrate the uniqueness of pressure as a physical parameter in the investigation of the structural and dynamic properties of aqueous surfactants. For many systems, the vibrational spectra at atmospheric pressure are very similar. Yet, the pressure tuning of these spectra will be able to provide additional information about the structure of the surfactant molecules and about their aggregation in water.

Literature Cited

1. Wong, P. T. T. In *Vibrational Spectra and Structure*, Durig, J. R., Ed.; Elsevier Press: Amsterdam, 1987, *Vol. 16*; pp 357-445.
2. Whalley, E.; Lavergne, A; Wong, P. T. T.   *Rev. Sci. Instrum.* **1976**, *42*, 845.
3. Wong, P. T. T.   *Chem. Phys. Lett.* **1981**, *77*, 291.
4. Wong, P. T. T.; Mantsch, H. H.; Snyder, R. G.   *J. Chem. Phys.* **1983**, *79*, 2369.
5. Wong, P. T. T.; Klug, D. D.   *Appl. Spectrosc.* **1983**, *37*, 284.
6. Wong, P. T. T.; Moffatt, D. J.   *Appl. Spectrosc.* **1983**, *37*, 85.
7. Wong, P. T. T.; Moffatt, D. J.   *Appl. Spectrosc.* **1984**, *38*, 599.
8. Wong, P. T. T.; Mantsch, H. H.   *J. Chem. Phys.* **1985**, *83*, 3268.
9. Wong, P. T. T.; Moffatt, D. J.; Baudais, F. L.   *Appl. Spectrosc.* **1985**, *39*, 733.
10. Wong, P. T. T.; Moffatt, D. J.   *Appl. Spectrosc.* **1989**, *43*, 1279.
11. Wong, P. T. T.; Moffatt, D. J.   *Appl. Spectrosc.* **1987**, *41*, 1070.

12. Mantsch, H. H.; Moffatt, D. J.; Casal, H. L.   *J. Mol. Spectrosc.* **1988**, *173*, 285.
13. Cameron, D. G.; Kauppinen, J; Moffatt, D. J.; Mantsch, H. H.   *Appl. Spectrosc.* **1982**, *36*, 245.
14. Cameron, D. G.; Umemura, J.; Wong, P. T. T.; Mantsch, H. H.   *Colloids and Surfaces,* **1982**, 4, 131.
15. Wong, P. T. T.; Mantsch, H. H.   *J. Colloid & Interface Science,* **1989**, *129*, 258.
16. Wong, P. T. T.; Mantsch, H. H.   *J. Chem. Phys.* **1983**, *78*, 7362.
17. Wong, P. T. T. In *High Pressure Chemistry and Biochemistry*, van Eldik, R.; Jonas, J. Eds.; Reidel Publ. Comp.: Dordrecht, 1987, NATO ASI Ser. C, *Vol.197*, pp. 381-400.
18. Wong, P. T. T.; Siminovitch, D. J.; Mantsch, H. H.   *Biochim. Biophys. Acta,* **1988**, *947*, 139.
19. Wong, P. T. T.; Mantsch, H. H.   *J. Phys. Chem.* **1983**, *87*, 2436.
20. Wong, P. T. T.   *Ann. Rev. Biophys. Bioeng.* **1984**, *13*, 1.

RECEIVED August 17, 1990

Chapter 4

Effect of Cholesterol on Location of Organic Molecules in Lipid Bilayers
An Infrared Spectroscopic Study

A. Muga and H. L. Casal[1]

Division of Chemistry, National Research Council of Canada, Ottawa, Ontario, K1A 0R6, Canada

An aliphatic ketone (9-heptadecanone) and two keto derivatives of stearic acid (as potassium salts) containing a ketone functionality either at position 5 or 12 were incorporated into bilayers of the phospholipid 1,2-dihexadecyl-sn-glycero-3-phosphocholine. Infrared spectra of these mixtures were measured as a function of temperature and amount of added cholesterol. It was found that the presence of cholesterol in these bilayers induces changes in the location of the guest ketone and that these changes are dependent on both temperature and cholesterol concentration. It is also demonstrated that, in the gel phase, the presence of cholesterol induces larger intersheadgroup separations and, therefore, water penetrates deeper into the lipid bilayer.

Cholesterol is an important and abundant constituent of most eukaryotic membranes. In cell membranes, cholesterol constitutes up to 50 mol % of the lipid. The physiological significance of cholesterol and its effect on the properties of lipid bilayers have been studied thoroughly. Most studies have dealt with cholesterol-induced changes in the phase behavior of lipids (1). There have also been studies of the effect of cholesterol in the rate of water permeation through membranes (2) and in the effect of this steroid on the insertion of metabolites into membranes (3, 4). Recently, it was shown that cholesterol modifies the pressure-induced expulsion of the local anesthetic tetracaine from lipid bilayers (5). It is accepted that cholesterol has a modulating effect on membrane properties; this modulating effect is summarized in the statement that cholesterol "fluidizes the gel phase of lipid bilayers while rigidifying their liquid-crystalline phase".

[1]Current address: BP Research, 4440 Warrensville Center Road, Cleveland, OH 44128

0097–6156/91/0447–0056$06.00/0
© 1991 American Chemical Society

In the work reported here, we examine the effect of cholesterol on the extent of hydrogen bonding between water and probe molecules. The probe molecules contain ketone functionalities and the extent of hydrogen bonding between these and water is determined directly from infrared spectroscopy. We have reported before (6-10) that this approach may be used in the characterization of different mesomorphic phases of surfactants and lipid bilayers. We find that the presence of cholesterol in bilayers of DHPC induces changes in the location of organic molecules within the bilayer and that these changes are dependent on both temperature and cholesterol concentration. We also find that, in the gel phase, the presence of cholesterol allows water to penetrate deeper into the lipid bilayer in accord with the recent findings of Hiff and Kevan (11).

Materials and Methods

1,2-Dihexadecyl-sn-glycero-3-phosphocholine (DHPC) was obtained from Fluka Chemical Corp. (Hauppauge, NY) and 9-heptadecanone (9HP) was from Aldrich (Milwaukee, WI); they were used as received. The preparation of 5-oxo potassium stearate (5-oxo KSA) and 12-oxo potassium stearate (12-oxo KSA) has been described before (10).

DHPC bilayers containing cholesterol and 9HP were prepared by codissolving in $CHCl_3$ the desired amounts and evaporating the solvent overnight under vacuum; the resulting lipid film was dispersed in excess D_2O and centrifuged. The lipid pellet was resuspended in D_2O to yield samples containing 0.15M DHPC. In all cases, the probe was present at concentrations of 0.01 M with the cholesterol concentration being 8, 29 and 45 mol % with respect to DHPC. For all the samples, the supernatant obtained after centrifugation was analyzed, the absence of the probe molecules in the supernatant shows that they are incorporated in DHPC. This is expected since their solubility in aqueous solution is practically nil. DHPC bilayers containing cholesterol and 5-oxo KSA or 12-oxo KSA were prepared as follows: DHPC and DHPC/cholesterol mixtures were dispersed in excess H_2O (1 mL) by gentle warming (50°C); micelles of 5-oxo KSA or 12-oxo KSA were prepared in H_2O at 70°C. These micelles were added to the ready-made DHPC dispersion (containing the desired amount of cholesterol) and this mixture was subjected to three cycles of heating to 70°C, vortexing while hot and cooling to 4°C. The diluted dispersion thus obtained was centrifuged at 14000 rpm for 15 min. in an Eppendorf centrifuge. The resulting lipid pellet was removed from the supernatant, dispersed in excess water and lyophilized for 14 hours. The finely divided powder thus obtained was redispersed in D_2O to yield a final DHPC concentration of 0.15 M in samples containing different amounts of cholesterol. For the incorporation of 5-oxo KSA and 12-oxo KSA in DHPC bilayers no alcohols were used as solvents (12) because alcohols may be incorporated in the membrane interior and could interact with the C=O groups of 5-oxo KSA and 12-oxo KSA.

Infrared spectra at 2 cm^{-1} resolution were recorded on a Digilab FTS-60 Fourier-transform spectrometer equipped with a DTGS detector. Typically, 200 interferograms were coadded for each spectrum. The samples were held in cells of 50-μm path length fitted with CaF_2 windows. For the hydration study (cf. Figure 2), a dry film was deposited on a CaF_2 window and this assembled as one window of a closed chamber, the atmosphere of this chamber may be controlled by circulating either dry nitrogen or nitrogen containing D_2O.

Results and Discussion

In this work, we examine the effect of cholesterol on the infrared spectra of DHPC bilayers containing a long-chain aliphatic ketone (9HP) or stearic acids (as potassium salts) with keto substituents at positions 5 or 12. By measuring the C = O stretching bands of these guest molecules it is possible to determine the extent of hydrogen bonding with the solvent (in this case D_2O). As shown recently (10) this yields location of the guest molecules in the bilayers. The approach is based on the sensitivity of the C = O stretching vibration of ketones to hydrogen bonding. The different species formed may be identified (13) as follows: a ketone C = O group not hydrogen bonded gives a C = O stretching band between 1718 and 1721 cm^{-1}; the C = O band of a monosolvate formed between a C=O group and water is between 1703 and 1706 cm^{-1}; the corresponding C = O band for a disolvate appears between 1695 and 1699 cm^{-1}. Therefore, it is possible to detect hydrogen bonding and to distinguish among the different species formed. The lipid studied, DHPC, is an ether-linked phospholipid; it was chosen because it does not give ester C = O stretching bands which would interfere with the bands of the guest molecules. We recall here the thermal phase behavior of DHPC since we study the temperature-induced changes in the C = O stretching bands as a function of cholesterol concentration. This phospholipid has been shown to form subgel, gel, ripple, and liquid-crystalline phases (14), the gel-to-liquid-crystal phase transition temperature, T_c, is 43.4°C and the pretransition is at 33.3°C. Below the pretransition and under high hydration conditions, the DHPC gel phase is characterized by interdigitated alkyl chains. Calorimetric experiments (7) have shown that the presence of long-chain ketones in DHPC bilayers results in a slight depression of the gel-to-liquid crystal transition temperature and more marked changes in the thermodynamic properties of the pretransition. It was concluded that long chain ketones are incorporated in the DHPC bilayers.

The temperature dependence of the C = O stretching bands in the infrared spectra of 9HP incorporated in DHPC bilayers, containing different amounts of cholesterol, is shown in Figure 1. For the sample containing no cholesterol (Figure 1A), the C = O groups of 9HP give a band at 1719 cm^{-1} at temperatures below T_c; upon heating above T_c, a band appears at ~1700 cm^{-1}, evident as a broad shoulder. The same pattern is observed in the sample containing 8 mol % cholesterol (Figure 1B). However, in this case the temperature at which the band at ~1700 cm^{-1} appears is shifted to lower values (between 35 and 39°C) compared with the cholesterol-free sample (Figure 1A).

At 29 mol % cholesterol below T_c there are three C = O bands at 1698, 1706 and 1719 cm^{-1} (Figure 1C). In this case, heating above T_c results in broadening of the 1719 cm^{-1} band and decrease in intensity of the other two bands; these remain present but only evident as an asymmetry to lower frequency on the strong 1719 cm^{-1} band. In the spectra of the sample containing 45 mol % cholesterol (Figure 1D) three bands also appear at the same frequencies as in Figure 1C. The difference is that the bands at 1698 and 1706 cm^{-1} (corresponding to hydrogen bonded C = O groups) are much more intense at temperatures below T_c. As in the case of the intermediate cholesterol concentration, heating above T_c induces significant decreases in their intensity such that they are only observed as a weak shoulder (Figure 1D).

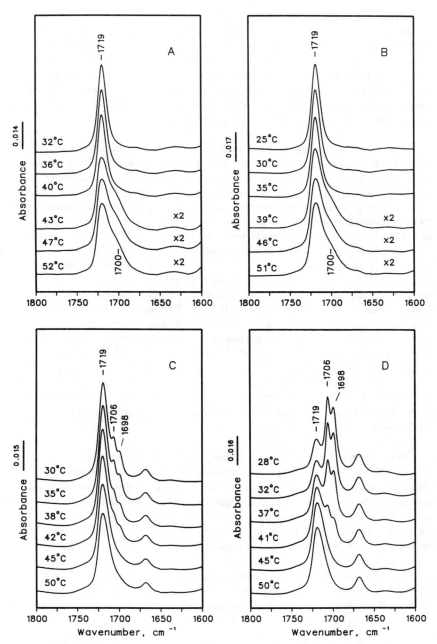

Figure 1: Infrared spectra (1800 to 1600 cm^{-1}) of 9-heptadecanone (9HP) incorporated in hydrated bilayers of DHPC containing 0 mol % (A), 8 mol % (B), 29 mol % (C) and 45 mol % cholesterol (D) at the indicated temperatures.

Taken together, the spectra shown in Figure 1 indicate that hydrogen bonding to the C = O group of 9HP is dependent on cholesterol concentration and temperature. Incorporation of cholesterol into DHPC bilayers at temperatures below T_c induces hydrogen bonding to the C = O group of 9HP demonstrated by increasing intensity of the bands at 1698 and 1706 cm^{-1}. In the liquid-crystalline phase, *i.e.*, at temperatures above T_c, the extent of hydrogen bonding is also dependent on cholesterol concentration; an increase in hydrogen bonding is observed for the samples with 0 and 8 mol % cholesterol (Figures 1A and 1B) while the opposite is found for the samples with 29 and 45 mol % cholesterol (Figures 1C and 1D).

The results presented so far indicate that the incorporation of cholesterol modifies the extent of hydrogen bonding between the C = O group of an aliphatic ketone (9HP) and a proton donor. The lipid molecule, DHPC, does not contain proton donors and therefore there is no possibility of lipid-induced hydrogen bonding to the guest ketone. Therefore, hydrogen bonding to this C= O group is either to the solvent (D$_2$O) or to the OH group at position 3 of cholesterol. To shed light on the question of which molecule serves as the proton donor we performed a series of control experiments. The sample containing 45 mol % cholesterol, chosen as the one inducing the largest changes, was spread as a film on a CaF$_2$ crystal and subjected to dehydration by circulating dry nitrogen at 28°C. The spectra recorded during dehydration are shown in Figure 2A. Clearly, as dehydration proceeds the intensity of the bands due to hydrogen bonded C = O groups (at 1698 and 1706 cm^{-1}) decreases. In the absence of D$_2$O there are no bands due to hydrogen bonded C = O groups. Furthermore, as shown in Figure 2B these changes are completely reversible; upon exposure of the film to an atmosphere containing D$_2$O, the bands at 1698 and 1706 cm^{-1} reappear. Also, in the infrared spectrum of 9HP dissolved in n-octane there is only one C = O stretching band at 1720 cm^{-1}, and addition of cholesterol such that its concentration is twelve times that of 9HP does not induce changes in the C = O stretching band of 9HP. From these control experiments, we conclude that cholesterol does not form hydrogen bonds directly with the C = O group of 9HP. We therefore interpret the effects observed on the C = O stretching bands of 9HP (Figure 1) as arising from 9HP - solvent (D$_2$O) contact.

An increase in hydrogen bonding to the 9HP molecule, evidenced from the bands at 1699 and 1706 cm^{-1}, may occur due to either water penetration into the lipid bilayer or to translocation of the guest ketone to a region where contact with the bulk water phase is favored. In order to distinguish between these two possibilities we studied systems containing C = O groups where translocation is either not possible or at least greatly diminished. This was achieved by incorporating two derivatives of potassium stearate, namely, 5-oxo potassium stearate (5-oxo KSA) and 12-oxo potassium stearate (12-oxo KSA). These molecules contain a carboxylate "head group" and are incorporated in lipid bilayers such that their COO- group is located near the lipid polar group and their chains embedded in the lipid hydrocarbon core (*17, 18*). Thus, the C = O groups at position 5 or 12 are 'fixed' allowing a relationship between membrane depth and solvent accessibility to be established.

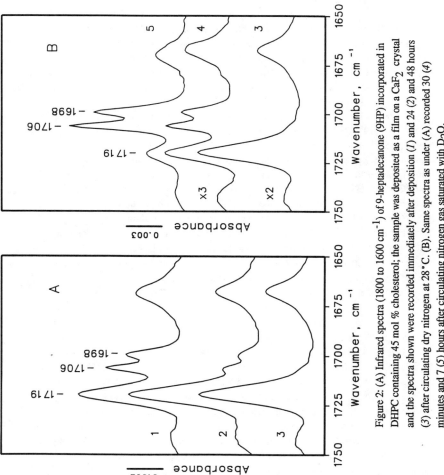

Figure 2: (A) Infrared spectra (1800 to 1600 cm^{-1}) of 9-heptadecanone (9HP) incorporated in DHPC containing 45 mol % cholesterol; the sample was deposited as a film on a CaF$_2$ crystal and the spectra shown were recorded immediately after deposition (*1*) and 24 (*2*) and 48 hours (*3*) after circulating dry nitrogen at 28 °C. (B). Same spectra as under (A) recorded 30 (*4*) minutes and 7 (*5*) hours after circulating nitrogen gas saturated with D$_2$O.

Mixtures of 5-oxo KSA and 12-oxo KSA with DHPC were prepared in the presence of 45 mol % cholesterol since it is at this concentration of cholesterol that the largest changes are observed. These experiments were performed at pH 7.5 and at this pH there is a measurable concentration of non-dissociated COOH groups in the derivatives of KSA. The pKa of stearic acid in lipid bilayers is 7.4 (19). The non-dissociated COOH groups give a broad band around 1700 cm^{-1} which interferes with the C = O stretching bands of the ketone groups (see Figures 3 and 4). Nevertheless, even with this interference valid information is obtained from the keto C=O stretching bands.

In Figure 3 we show the C = O stretching bands of 5-oxo KSA incorporated in DHPC bilayers containing 45 mol % cholesterol and in the absence of cholesterol. In the spectrum recorded at 22°C (Figure 3A) there are two bands at 1706 and 1718 cm^{-1} corresponding to hydrogen bonded and non-hydrogen bonded C = O groups respectively. Addition of 45 mol % cholesterol results in an increase of the intensity of the band at 1706 cm^{-1}. At 53°C (Figure 3B) in the absence of cholesterol, the C = O stretching band is centered at 1709 cm^{-1}. This shift to lower frequencies is due to an increase in the intensity, of the band at 1706 cm^{-1} relative to that of the 1719 cm^{-1} band; addition of cholesterol at this temperature results in an increase of the 1719 cm^{-1} band intensity evidenced as broadening at this frequency. Thus, the presence of cholesterol increases the hydrogen bonding between D$_2$O and the keto C = O group at temperatures below T$_c$, while decreasing it above this temperature. This behavior parallels the so-called "dual-effect" of cholesterol on lipid bilayers: cholesterol fluidizes bilayers at temperatures below T$_c$ and rigidifies them above T$_c$ with respect to the cholesterol-free system (1). We interpret the cholesterol-induced increase in hydrogen bonding below T$_c$ (Figure 3A) as being due to an 'opening' of the bilayer which allows more solvent to penetrate. Similar conclusions were reached from studies of doxyl stearic acid spin probes in frozen lipid vesicles (11) and agree with the notion that cholesterol separates the polar groups and increases the size of lipid vesicles (20-21). Our results provide further information, that at temperatures above T$_c$ the presence of cholesterol decreases hydrogen bonding to the C = O group at position 5.

The C = O stretching bands in the spectra of 12-oxo KSA incorporated in DHPC bilayers are shown in Figure 4 as a function of temperature and in the presence (45 mol%) and absence of cholesterol. In these spectra, the effects due to heating and cholesterol are less marked than for 5-oxo KSA. At 22°C, i.e., below T$_c$, the C = O stretching band of 12-oxo KSA is at 1719 cm^{-1} (Figure 4A), and the same is observed in the presence of 45 mol % cholesterol. At 53°C, i.e., above T$_c$, the C = O band is also at 1719 cm^{-1} (Figure 4B).; the width of this C= O stretching band is larger at temperatures above T$_c$, but the presence of cholesterol induces narrowing with respect to the cholesterol-free system. The broadening of the C = O stretching band observed above T$_c$ in the cholesterol-free sample is due to motional effects as the bilayer is more disordered above T$_c$. The presence of cholesterol restricts the motional disorder of the chains and therefore broadening does not occur to the same extent.

The gel to liquid crystal phase transition temperature of DHPC bilayers containing 5-oxo KSA or 12-oxo KSA, at the concentrations used here, is around 46°C (data not shown),

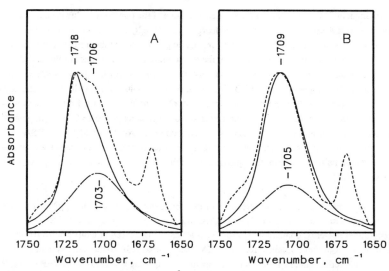

Figure 3: Infrared spectra (1800 to 1600 cm^{-1}) of 5-oxo potassium stearate (5-oxo KSA) incorporated in hydrated DHPC bilayers in the absence (—) and presence of 45 mol % cholesterol (—·); (A) spectra recorded at 22 °C, and (B) spectra recorded at 53 °C. The broad bands at 1703 cm^{-1} (A) and 1705 cm^{-1} (B) are due to the COOH group of stearic acid (see Text).

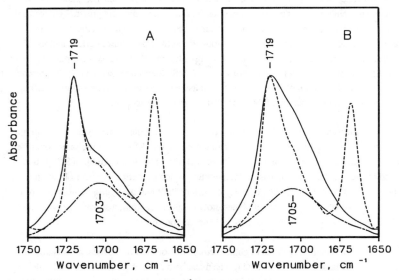

Figure 4: Infrared spectra (1800 to 1600 cm^{-1}) of 12-oxo potassium stearate (12-oxo KSA) incorporated in hydrated DHPC bilayers in the absence (----) and presence of 45 mol % cholesterol (—·-); (A) spectra recorded at 22 °C, and (B) spectra recorded at 53 °C. The broad bands at 1703 cm^{-1} (A) and 1705 cm^{-1} (B) are due to the COOH group of stearic acid (see Text).

approximately 3°C more than for aqueous suspensions of the pure lipid. This is an indirect indication that these guest molecules are incorporated in the hydrocarbon region of DHPC bilayers, and is in accord with calorimetric results on phospholipid bilayers containing fatty acids (22). It is important to point out that the temperature-induced changes of the C = O stretching band for both keto-surfactants incorporated in cholesterol-free DHPC bilayers occur within a narrow temperature range (~3°C), which coincides with the gel to liquid crystal phase transition of the lipid. The presence of 45 mol % cholesterol abolishes the lipid phase transition and decreases the magnitude of the temperature-induced changes of the C=O stretching bands. The different cholesterol-induced changes in the spectra of 5-oxo KSA and 12-oxo KSA incorporated in DHPC bilayers, demonstrate that the effect of cholesterol is dependent on the bilayer depth; being larger near the lipid-water interface, in accord with recent results of Hiff and Kevan (11).

We now return to the interpretation of the results obtained with 9HP as guest in light of the findings with 5-oxo KSA and 12-oxo KSA. It is evident that the spectra of 9HP incorporated in DHPC containing 45 mol % cholesterol (Figure 1D) at temperatures below T_c are different from those obtained for 5-oxo KSA and 12-oxo KSA in the same system (Figures 3 and 4). This reflects the fact that, in this temperature range, the C = O group of 9HP is in a different environment than the C = O groups of 5-oxo KSA and 12-oxo KSA. Some characteristics of the environment may be derived from the spectral features of the C = O stretching bands. Firstly, the C = O group is closer to the aqueous interface than the C = O group of 5-oxo KSA; the strong bands at 1706 and 1698 cm^{-1} in the spectrum of 9HP indicate direct contact with the solvent. Secondly, these bands are narrow, showing that increasing cholesterol concentrations below T_c induces a restricted motional state. The same spectral characteristics have been described for guest ketones in temperature and pressure induced gel phases (8, 9) where a bent conformation has been proposed for the guest ketones. In this conformation the C = O group of 9HP is immobilized at the lipid-water interface and its alkyl chains are embedded in the first segments of the lipid hydrocarbon chains corresponding to the so-called "plateau region" of molecular order. ^{2}H NMR studies have shown that the presence of cholesterol (at 29 and 45 mol %) increases the molecular order of this chain segment (23) in accord with the immobilization of the guest 9HP molecule observed here.

At temperatures above T_c the spectra of 9HP at high cholesterol concentration (Figure 1D) resemble those of 12-oxo KSA rather than those of 5-oxo KSA under the same conditions. This demonstrates that hydrogen bonding between D$_2$O and the C = O group of 9HP is not favored above T_c. Thus, an increase in temperature, which decreases the order of the lipid alkyl chains, allows a deeper penetration of the guest ketone 9HP into the hydrophobic region of the bilayer. In summary, the presence of cholesterol at high concentration above T_c favors the solubilization of the guest 9HP in the bilayer hydrocarbon core.

A similar modulating effect of cholesterol on the location of the local anesthetic tetracaine has been observed recently (5). The uncharged form of the tetracaine molecule intercalates deeply into DHPC bilayers in the gel state and is displaced to an environment close to the lipid-water interface by increasing the hydrostatic pressure, in other words, by increasing the alkyl chain order, only in the presence of cholesterol (30 mol %). Moreover, it was found, as we do here, that the cholesterol-induced migration is a completely reversible process.

In an attempt to compare the physical state of the bilayers with the extent of hydrogen bonding to the keto group of 9HP, as a function of cholesterol concentration, we studied the temperature dependence of the CH_2 symmetric stretching vibrational modes, ν_s (CH_2), (24) (Figure 5B) and the thermal behavior of the empirical parameter I_b/I_f (Figure 5A). The methylene symmetric stretching vibrational mode is sensitive to the physical state of the bilayer; its frequency increases when disorder, in the form of gauche bonds, is introduced into hydrocarbon chains. The parameter I_b/I_f, as described before (6, 7) represents the ratio of the intensity of the hydrogen bonded C = O stretching bands (I_b) to the intensity of the non-bonded band (I_f).

For the samples containing no cholesterol and 8 mol % cholesterol, I_b/I_f increases abruptly at temperatures coinciding with the gel to liquid crystal phase transition. The opposite temperature dependence of I_b/I_f is observed in the sample with the intermediate cholesterol concentration (29 mol %); in this case, there is a small increase in I_b/I_f up to 38 °C and at higher temperatures it decreases to practically zero (Figure 5A). The presence of 29 mol % cholesterol, as indicated in Figure 5B, reduces the frequency shift of the ν_s (CH_2) mode, indicating that although greatly diminished, there still remains a phase transition in this sample. This is in accord with the generally accepted idea that the presence of cholesterol at concentrations higher than 22 mol %, reduces the temperature induced physical changes of bilayers containing this steroid (26-30). Also, for this cholesterol concentration (29 mol %) both parameters ν_s (CH_2) and I_b/I_f, change at the same temperature.

The thermal dependence of I_b/I_f for the sample containing 45 mol % cholesterol is qualitatively similar to that of the sample with the intermediate cholesterol concentration (Figure 5A). Increasing the temperature from 10 to 30 °C results in a modest increase in I_b/I_f; at temperatures above 30 °C, I_b/I_f decreases such that its value is practically zero at 45 °C. Thus, even though the temperature dependence of the (CH_2) mode in the sample containing 45 mol % cholesterol suggests that the phase transition is completely abolished, there is a marked temperature dependence of the parameter I_b/I_f. Therefore, there are temperature-induced changes in the bilayers inducing translocation of the guest 9HP which are not reflected in the (CH_2) mode frequency. These effects are due to differences in the packing of the lipid hydrocarbon chains and evident in the characteristics of the CH_2 scissoring vibrations.

The CH_2 scissoring bands of the DHPC alkyl chains are located in the 1480 - 1460 cm^{-1} spectral region and are characteristic of the nature of the chain packing in the gel phase (24). Above the phase transition temperature of DHPC, for DHPC: 9HP (12:1 molar ratio) bilayers containing the cholesterol concentrations used in this study, a single CH_2 scissoring band is observed at ~1468 cm^{-1} (data not shown), representative of alkyl chains packed without interchain coupling (31). Lowering of the temperature induces broadening of this band, due to the appearance of two bands as a consequence of interchain coupling. These two bands are clearly seen in the difference spectra (for details see reference 31). Such difference spectra

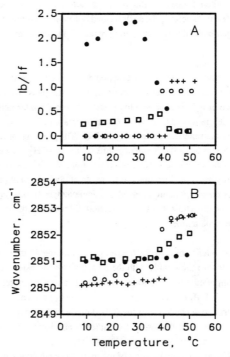

Figure 5: (A) Temperature dependence of the parameter I_b/I_f obtained from the spectra of 9-heptadecanone (9HP) incorporated in hydrated DHPC bilayers containing different cholesterol concentrations: 0 mol % (+); 8 mol % (o); 29 mol % (◻) and 45 mol % (•) (B) Temperature dependence of the symmetric CH_2 stretching band of DHPC in the same samples described in (A).

determined from the samples measured here are shown in Figure 6. It is evident that at 29 and 45 mol % cholesterol, the splitting of the CH_2 scissoring band at 20°C is less marked than for the samples without cholesterol and with 8 mol % cholesterol. In the absence of cholesterol, we believe that the presence of 9HP at a concentration of about 8 mol % of the total lipid, does not prevent the formation of an interdigitated lamellar gel phase. The observation of a similar splitting in the gel phase of pure DHPC bilayers supports this hypothesis. Moreover, it has been demonstrated (32) that incorporation of 20 mol % of the uncharged form of tetracaine in DHPC bilayers does not remove the interdigitation of its lamellar gel phase. X-ray diffraction studies of hydrated mixtures of DPPC and DHPC (15) have also shown that when less than 50 mol % DPPC is added, DHPC remains in the interdigitated gel phase and shows full solubility for DPPC up to equimolarity without major structural changes. Comparing the difference spectra of the sample without cholesterol and with 8 mol % cholesterol, it could be suggested that, in both cases, DHPC interdigitates in the gel phase.

The effects of the incorporation of an equimolar concentration of cholesterol on the lipid packing, conformation and dynamics of DHPC bilayers have been studied by X-ray diffraction and ^{2}H-NMR spectroscopy (33). The results of these studies indicate that addition of 50 mol % cholesterol to DHPC prevents the formation of an interdigitated phase at 22°C. The difference spectra in the methylene scissoring region for the sample containing 45 mol % cholesterol, reveal that at 20°C the splitting is much less pronounced that for the samples containing no cholesterol and 8 mol % cholesterol. The reduction of the splitting is due to the prevention of the interdigitation at this cholesterol concentration (45 mol %). For the intermediate cholesterol concentration (29 mol %) the difference spectra in the gel phase resemble those of the highest one. However, the splitting is slightly more marked in the former, suggesting that the presence of this cholesterol concentration does prevent, to a certain extent, the interdigitation of the DHPC alkyl chains.

Thus, the partial or total removal of the interdigitation of the lipid alkyl chains by cholesterol seems to be an important factor in determining the ability of the ketone molecule to place the $C = O$ group close to the polar interface, at low temperatures. As the interdigitation is removed, the contact between the $C = O$ group of the 9HP molecule and the solvent increases.

The difference spectra of the sample without cholesterol show that the interchain coupling (reflected in the two band minima) is removed between 40 and 43°C coinciding with T_c. Above T^c, X-ray diffraction studies (33) have demonstrated the lack of interdigitation in DHPC bilayers. It is also at this temperature, that the parameter I_b/I_f increases demonstrating a direct relationship between the physical state of the bilayer and the properties of the guest ketone. A similar observation is made for the sample containing 8 mol % cholesterol.

In the sample containing 45 mol % cholesterol the DHPC alkyl chains are not interdigitated (33). However, the difference spectra (Figure 6D) show that up to 25°C there are two CH_2 scissoring bands reflecting interchain coupling, and that it is not observed at

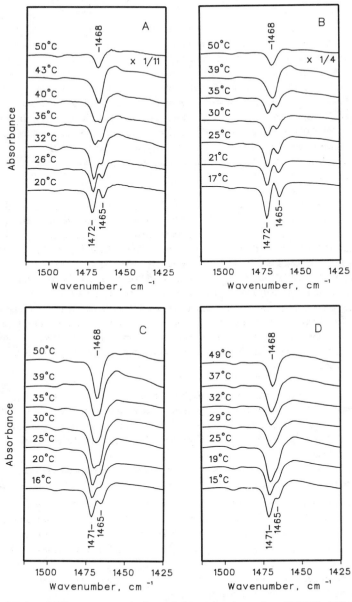

Figure 6: Infrared difference spectra in the methylene scissoring region of DHPC : 9HP (12 : 1 molar ratio) bilayers containing (A) 0 mol % cholesterol (B) 8 mol %, (C) 29 mol %, and (D) 45 mol % cholesterol. The corresponding temperatures are the higher values of the ~4.0°C intervals in which subtractions have been performed.

temperatures above 28°C. This coincides with the temperature at which the parameter I_b/I_f decreases for this sample (Figure 5A). Thus, there is a direct correlation between interchain vibrational coupling and I_b/I_f. The same relationship is observed in the sample containing 29 mol %, showing that the ketone may translocate only when the alkyl chains are motionally disordered such that there is no vibrational coupling.

Conclusion

From the data presented here several conclusions may be reached regarding the effect of cholesterol on lipid bilayers. It is shown that, even if the presence of cholesterol in bilayers serves to moderate temperature-induced changes, its ability to affect the location of solubilized molecules is highly temperature dependent. We have also shown, in accord with previous work (11), that the presence of cholesterol in the gel phase results in a larger separation between the lipid polar groups and this in turn allows water to penetrate into the lipid hydrophobic core. This penetration is apparent only at short distances from the lipid-water interface.

Recent experiments have shown that the non-specific, physical chemical interactions between small hydrophobic, water-insoluble molecules and the hydrocarbon chains of lipid membranes are important determinants of the rate at which these molecules enter cells and are metabolized (3, 34). Cholesterol has the capability of modifying these interactions and also increases the affinity of vesicle surfaces for amphiphillic molecules (4) separating the lipid polar groups (35).

Cholesterol can modify both the hydrophobic attraction between lipid hydrocarbon chains and electrostatic interactions between lipid polar groups. The influence it has on the location of 9HP reflects this dual effect. At low temperature, the "spacer" effect of cholesterol allows the ketone to gain access directly to the lipid-water interface. At high temperatures, a more disordered hydrocarbon core favors the solubilization of the guest molecule.

This effect of cholesterol on the location of guest molecules has been cited for chlorophyll a (36) and tetracaine (5) both molecules with specific biological function. In the present work we have been able to observe this effect over the physiological ranges of temperature and cholesterol concentration. Future experiments with other molecules would clarify if this property of cholesterol is generally applicable to other systems and, furthermore, if it extends to interactions between two molecules solubilized in a membrane.

Legend of Symbols

DHPC: 1,2-dihexadecyl-sn-glycero-3-phosphocholine; KSA: potassium stearate; 5-oxo KSA: 5-oxo potassium stearate; 12-oxo KSA: 12-oxo potassium stearate; 9HP:9-heptadecanone; T_c: temperature of the gel-to-liquid crystal phase transition of DHPC bilayers; 43.4!C

Literature Cited

1. Yeagle, P.L. (1985) *Biochim. Biophys. Acta.*, **822**, 267-287.
2. Lawaczeck, R. 91979) *J. Membrane Biol.*, **51**, 229-261.
3. Cooper, R.B., Noy, N., & Zakim, D. (1987) *Biochemistry*, **26**, 5890-5896.
4. Fukushima, D., Yokoyama, S., Kzdy, F.J., & Kaiser, J. (1981) *Proc. Natl. Acad. Sci. USA*, **78**, 2732-2736.
5. Auger, M., Jarrell, H.C., Smith, I.C.P., Wong, P.T.T., Siminovitch, D.J. and Mantsch, H.H. (1987) *Biochemistry*, **26**, 8513-8516.
6. Casal, H.L. (1988) *J. Am. Chem. Soc.*, **110**, 5203-5205.
7. Casal, H.L. (1989) *J. Phys. Chem.*, **93**, 4328-4330.
8. Casal, H.L. & Martin, A. (1989) *Can. J. Chem.*, **67**, 1554-1559.
9. Casal, H.L. & Wong, P.T.T. (1990) *J. Phys. Chem.*, **94**, 777-780.
10. Muga, A. & Casal, H.L. (1990) *J. Phys. Chem.* in press.
11. Hiff, T., & Kevan, L. (1989) *J. Phys. Chem.*, **93**, 1572-1575.
12. Eisinger, J., & Flores, J. (1983) *Biophys. J.*, **41**, 367-379.
13. Symons, M.C.R. & Eaton, G. (1985) *J. Chem. Soc.*, Faraday Trans. 1, **81**, 1693-1977.
14. Ruocco, M.J., Siminovitch, D.J. & Griffin. R.G. (1985) Biochemistry 24, 2406-2411.
15. Kim, J.T., Mattai, J., & Shipley, G.G. (1987) *Biochemistry*, **26**, 6592-6598.
16. Laggner, P., Lohner, K., Degovics, G., Mller, K. & Schuster, A. (1987) *Chem. Phys. Lipids*, **44**, 31-60.
17. Ramachandran, C., Pyter, R.A., & Mukerjee, P. (1982) *J. Phys Chem.*, **86**, 3198-3205.
18. Baglioni, P., Ferroni, E., Martini, G., & Ottaviani, M.F., (1984) *J. Pys. Chem.*, **88**, 5107-5113.
19. Ptak, M.; Egret-Charlier, M.; Sanson, A.; Boulousa, O. *Biochim, Biophys. Acta*, 1980, **600**, 387-397.
20. de Kruyff, B., Cullis, P.R., & Redda, G.K. (1976) *Biochim. Biophys. Acta*, **436**, 729-740.
21. Johnson, S.M. (1973) *Biochim. Biophys. Acta.*, **307**, 27-41.
22. Schullery, S.E., Seder, T.A. Weinstein, D.A. and Bryant, D.A. (1981) *Biochemistry*, **20**.
23. Dufourc, E.J. Parish, E.J., Chitrakorn, S., & Smith, I.C.P. (1984) *Biochemistry*, **23**, 6062-6071.
24. Casal, H.L. and Mantsch, H.H. (1984) *Biochim. Biophys. Acta*, **779**, 381-401.
26. Knoll, W., Schmidt, G., Ibel, K., & Sackmann, E. (1985) *Biochemistry*, **24**, 5240-5246.
27. El-Sayed, M.Y., Guion, T.A., & Fayer, M.D. (1986) *Biochemistry*, **25**, 4825-4832.
28. Carruthers, A., & Melchior, D.L. (1983) *Biochemistry*, **22**, 5797-5807.
29. Blok, M.C., Van Deenen, L.L.M., & de Gier, J. (1977) *Biochim. Biophys. Acta*, **464**, 509-518.
30. Presti, F.T., Pace, R.J., & Chan, S.I. (1982) *Biochemistry*, **21**, 3831-3835.
31. Umemura, J., Cameron, D.G. and Mantsch, H.H. (1980) *Biochim. Biophys. Acta*, **602**, 32-44.
32. Auger, M.; Jarrell, H.C.; Smith, I.C.P.; Siminovitch, D.J.; Mantsch, H.H., & Wong, P.T.T. (1988) *Biochemistry*, **27**, 6086-6093.
33. Siminovitch, D.J., Ruocco, M.J., Makriyannis, A., & Griffin, R.G. (1987) *Biochim. Biophys. Acta.*, **901**, 191-200.
34. Noy, N., Donnelly, T., & Zakim, D. (1986) *Biochemistry*, **25**, 2013-2021.
35. McIntosh, T.J., Magid, A.D.; & Simon, S.A. (1989) *Biochemistry*, **28**, 17-25.
36. Ford, W.E. & Tollin, G. (1984) *Photochem. Photobiol.*, **40**, 249-259.

RECEIVED August 2, 1990

Chapter 5

Phase Behavior in Aqueous Surfactant Systems
Infrared Studies

Curtis Marcott[1], Robert G. Laughlin[1], Andre J. Sommer[2], and J. E. Katon[2]

[1]Miami Valley Laboratories, The Proctor and Gamble Company, P.O. Box 398707, Cincinnati, OH 45239–8707
[2]Department of Chemistry, Molecular Microspectroscopy Laboratory, Miami University, Oxford, OH 45056

A new technique for studying the phase behavior of aqueous surfactant systems using infrared (IR) microspectroscopy has been developed. All the phases which exist along a particular isotherm are formed by creating an interface between water and surfactant within a long (55 mm) 25-μm pathlength fused silica cell, and allowing the components to diffuse together. Analyses of the entire composition range can be achieved using near–IR spectra. The analyses can be based on either the hydrogen-stretching fundamentals or the overtone and combination bands, depending on the percentage of water present. IR data may in some cases also serve to identify the phases present. Exploratory studies of the decyldimethylphosphine oxide/water system have been performed which demonstrate the validity of this technique.

The phase diagrams of aqueous surfactant systems provide information on the physical science of these systems which is both useful industrially and interesting academically (1). Phase information is thermodynamic in nature. It describes the range of system variables (composition, temperature, and pressure) wherein smooth variations occur in the thermodynamic density variables (enthalpy, free energy, etc.), for macroscopic systems at equilibrium. The boundaries in phase diagrams signify the loci of system variables where discontinuities in these thermodynamic variables exist (2).

The phase behavior of surfactant systems is particularly complex because of the existence of numerous lyotropic (solvent-induced) liquid crystal phases (3). These phases, like liquids and crystals, are discrete states of matter. They are fluids, but their x-ray patterns display sharp lines signifying the existence of considerable structure. They are often extremely viscous; because of their high viscosities and for other reasons they are difficult to study using conventional methods. This is evident from the fact that serious errors in the presumably well-established classical aqueous phase diagrams of soaps, sodium alkyl sulfates, monoglycerides, and

0097–6156/91/0447–0071$06.00/0
© 1991 American Chemical Society

polyoxyethylene surfactants have recently been discovered ($\underline{4}$,$\underline{5}$, Laughlin, R. G. Am. Inst. Chem. Eng. Symposium Series, in press.). Aqueous surfactant systems have been investigated largely using isoplethal methods wherein temperature is varied at constant composition, but qualitative swelling methods have played a prominent role in recent studies ($\underline{6}$). The "penetration" or "flooding" experiment, for example, is useful as a means of documenting the sequence of phases present along a particular isotherm. The structures of phases produced by swelling have been determined using both optical texture and synchrotron x-ray studies ($\underline{5}$).

A quantitative phase studies method based on the swelling principle has also been developed; it is termed the diffusive interfacial transport (DIT) method ($\underline{7}$). In the DIT method, phases are formed by swelling the surfactant within a long, thin, silica capillary having a chamber with a rectangular cross-section and optically flat walls. The swelling produces phase bands separated by interfaces; composition profiles within these bands extrapolated to the position of the interface provide quantitative information as to the compositions of coexisting phases.

In the diffusive interfacial transport-refractive index (DIT-NDX) method, compositions are determined using precise refractive index data ($\underline{8}$). Refractive index data valid to +/- 0.00005 are obtainable using the DIT apparatus within an area of 30 μm^2 in a sample approximately 25-μm thick (0.75 picoliter volume). Data collection and analysis require 9 seconds. The accuracy, spatial resolution, and speed with which refractive indices can be determined are thus superb.

Refractive index data are very useful for the quantitation of isotropic (liquid and cubic liquid crystal) phases, and for the calibration of cell thickness and nonflatness. However, the analysis of birefringent phases using refractive index data has been found to be unreliable ($\underline{9}$). A problem arises from the fact that the orientation of such phases relative to the direction of the light path, as well as the system variables, influence refractive indices. In order to use refractive index data for quantitation, a phase must spontaneously orient in a reproducible fashion. Such orientation does occur in the case of fluid lamellar phases (as in short chain polyoxyethylene nonionic systems ($\underline{7}$)), but viscous lamellar phases, hexagonal phases, and crystal phases do not orient to a sufficient degree.

Thus, while the value of executing phase studies using swelling methods has been established beyond doubt, the analysis of composition using index-of-refraction data has been disappointing. A different kind of data which is useful in the context of the DIT experiment and which could serve as an alternative basis for analysis is needed.

The recent advent of infrared microspectroscopy ($\underline{10}$) suggested the possibility that infrared absorbance data could serve as the basis for analysis. It seemed unlikely that the absorbance of water in a liquid crystal phase would be anisotropic to the same degree as is refractive index. Further, infrared data could possibly provide qualitative information as to phase structure – a kind of information not contained in refractive index data ($\underline{11}$).

During the present work the decyldimethylphosphine oxide ($C_{10}PO$)/water system was used as a model system to explore the feasibility of using infrared data for analysis. Information on the phase diagram of this system was available, though incomplete (see Figure 1) (12). The L in Figure 1 represents the isotropic liquid phase, E the hexagonal phase, D the lamellar liquid crystal phase, and X the dry crystal. This surfactant displays a classical phase sequence at room temperature. It is thermally stable and melts reversibly at a moderate temperature, so that the studies are easily performed. Swelling rates are such that data may conveniently be collected during the next day or two after initiation.

Experimental

All infrared spectra were recorded with an IR-PLAN microscope (IR-PLAN is a registered trade mark of Spectra Tech, Inc.) integrated to a Perkin-Elmer Model 1800 Fourier transform infrared (FT-IR) spectrophotometer. The spectrophotometer consisted of a proprietary heated wire source operated at 1050°C, a germanium overcoated potassium bromide beamsplitter, and a narrow-band mercury-cadmium-telluride (HgCdTe) detector. The detector was dedicated to the microscope and had an active area of 250 x 250 µm. The entire optical path of the system microscope was purged with dry nitrogen.

Double-sided interferograms were collected at 8 cm^{-1} resolution using mirror velocities of 1.5 or 3.0 cm/s. Either 512 or 128 scans were co-added depending on the signal-to-noise level desired in the final spectra. Reference interferograms were sampled at every other zero crossing, yielding a spectral range of 7900 to 0 cm^{-1}. Interferograms were self corrected for phase errors using 128 points on either side of the centerburst. Transformation of these interferograms into spectra was performed using a fast Fourier transform and a medium Beer-Norton apodization function. Single-beam spectra of the background and sample were then ratioed to yield absorbance spectra covering the range from 7900 to 1800 cm^{-1}.

A diagram of the silica DIT cell used in both the refractive index and infrared experiments is shown in Figure 2. The overall thickness of the DIT cell is 2.8 mm. The cell is 6.5-mm wide by 55-mm long. The 25-µm pathlength sample chamber is 450-µm wide and runs the entire length of the cell. The long direction of the cell is assigned the x, the width the y, and the direction normal to the chamber the z direction. Light rays pass through the cell during optical studies in the z direction.

The sample thickness in the DIT cells (25 µm) is very large in comparison with the usual thickness of samples for infrared transmission studies, and the cell itself is also thick (2.8 mm). Studies in the attenuated total reflectance (ATR) mode do not fit the experiment. For all these reasons, it was by no means certain that infrared studies could in fact be performed using the existing DIT cells.

Optical Considerations. The objective of this investigation was to determine the amount of water in coexisting phases of the $C_{10}PO$ system. The value of the data is therefore determined by the instrument's photometric accuracy. Several factors which contribute to photometric inaccuracies in infrared microspectroscopy include sample geometry, spherical aberrations caused by the introduction of

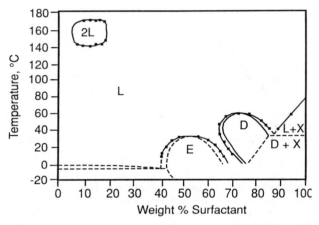

Figure 1.   Phase   diagram   of   the   decyldimethylphosphine
oxide/water system.

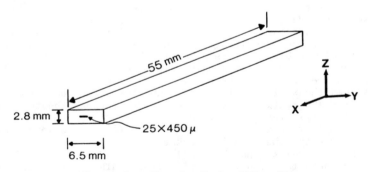

Figure 2.   Sketch of the DIT cell.

the sample into the optical path, and diffraction. Three key features of the IR-PLAN infrared microscope which facilitated these analyses included a variable focus, an objective and condenser which could compensate for spherical aberrations introduced by thick samples, and dual remote image masking. The impact of each of these characteristics on photometric accuracy will now be discussed.

The IR-PLAN microscope consisted of a 15X objective with a numerical aperature (N.A.) of 0.58, and a 10X (N.A. = 0.71) condenser. An optical diagram of the microscope is presented in Figure 3. The samples were viewed with either bright field or polarized light illumination using white light. The 10X binocular and 15X objective provided a total magnification of 150X. The adjustable focus of the microscope permitted the remote apertures to be imaged in the center of the relatively thick (2.8 mm) DIT cell. Spherical aberrations introduced by the DIT cell were compensated by decreasing the distance between the primary and secondary mirrors of the objective and condenser. This adjustment is necessary to increase the contrast of the aperture edges (i.e. optimize spatial definition and spatial resolution) when imaging samples imbedded in a thick matrix. The adjustment is made with respect to normal viewing conditions and ultimately enhances the throughput of the microscope under the given conditions. The optimum decrease in separation between the mirrors for each optical element was determined using equations provided by Spectra Tech, Inc. (13) in conjunction with the known thickness of the cell and its refractive index. The decrease in separation between the primary and secondary mirrors of the 15X objective and the 10X condenser were estimated to be 0.273 mm (21 divisions) and 0.546 mm (42 divisions), respectively.

Diffraction is defined as the ability of light to bend around samples or high contrast edges (i.e., apertures), and is presently the largest contributor of stray light in IR microspectroscopy. Diffraction can cause impurities from the neighboring matrix to be observed in the spectra of small samples embedded in matrices. In the case of small isolated samples, diffraction can cause nonlinearities in the transmittance values at longer wavelengths. Detailed discussions of the effects of diffraction with regard to photometric accuracies have been recently published (14–16). The dual remote image masking employed by the IR-PLAN significantly reduces photometric inaccuracies caused by diffraction.

The geometry of the chamber of the DIT cell is similar to a cross-section of a laminate whose individual layers are relatively large. For this configuration the effects of diffraction will, at most, decrease the spatial resolution of the technique. These effects, however, will not become significant if the spatial resolution is above 10 µm. Aperture sizes employed in this investigation were 50 x 160 µm with the smaller dimension corresponding to the spatial resolution in the x direction.

The sample geometry is seemingly ideal, but the cell and its pathlength must be considered for proper quantitation. A major difference between quantitation on macro samples and micro samples is the fact that the pathlength for a sample of given thickness is much larger in the micro experiment. This increased pathlength is due to the high angles at which the rays pass through the sample with respect to the sample normal. Figure 4 illustrates the pathlength through the cell for peripheral rays exiting the objective (top) and

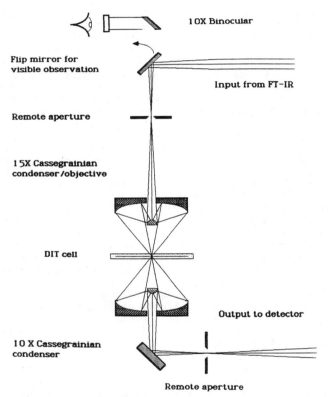

Figure 3.   Diagram of the Spectra Tech IR-PLAN FT-IR microscope.

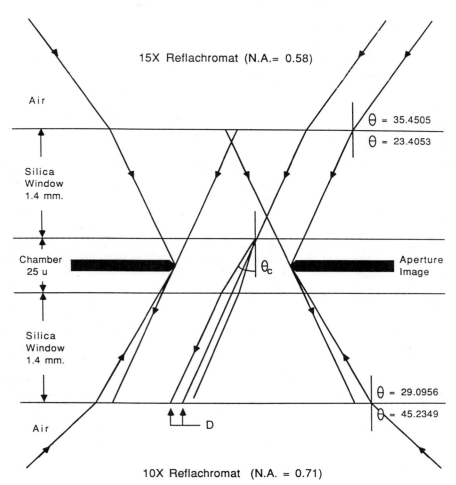

Figure 4. Optical ray diagram through the DIT cell.

those being collected by the condenser (bottom). Although the higher
N.A. of the condenser would permit rays with a larger angle to be
collected, it can be seen that the objective (i.e. the optical
element with the lowest numerical aperture) limits the pathlength for
this configuration.

Other factors of equal importance with regard to pathlength are
the aperture size, the change in refractive indices experienced by
rays passing through the different interfaces within the cell, and
changes in refractive indices associated with changing sample
composition along the x direction. Changing the aperture size will
change the pathlength; as a result, it is imperative that precisely
identical apertures be employed for the collection of the reference
and sample spectra. The change in refractive indices experienced by
rays passing through different interfaces within the cell could not
be controlled for the sample spectra but could be controlled for the
collection of the reference spectra by a careful selection of
materials.

Early investigations indicated that the choice of a proper
material used to generate the reference spectrum was extremely
important. Using the entire thickness of the silica cell (i.e.
focusing through the cell next to the chamber) as a reference
material was unacceptable because of differences between the
transmittance of the silica and the sample. Likewise, using the
empty cell to obtain the reference spectrum was not acceptable,
because the position of the aperture image is completely different
from that of the sample and serious problems with interference
fringes result. The interference fringes arose from reflections of
the light rays at interfaces with large differences in refractive
indices. In an effort to match the refractive index of the solution
to that of the cell, Freon 11 (fluorotrichloromethane, refractive
index = 1.49) and carbon tetrachloride (refractive index = 1.46) were
employed as the reference material. The cell filled with Freon also
produced interference fringes, but when filled with carbon
tetrachloride, it yielded reference spectra free of any defects.

Figure 4 depicts the angle, Θ_c, to which the extreme rays are
refracted when passing from the silica window to the cell chamber.
When the chamber is filled with a material that matches the
refractive index of the silica (e.g. carbon tetrachloride, n = 1.46),
the ray does not deviate from its path. If the material filling the
chamber has a refractive index larger than the silica (e.g. freon,
n = 1.49), the ray is bent toward the surface normal. Conversely,
materials with indices lower than silica will refract the ray away
from the surface normal. The total deviation "D" of the ray for
materials ranging in refractive index from 1.33 to 1.60 is
approximately 2.4 µm. Considering the angles at which the rays enter
the chamber, the corresponding deviation in the z direction is ± 3µm.
This variation will have the effect of changing the focus (i.e. the
coordinate at which the aperture image appears) of the microscope on
the z axis.

Perhaps more important is the fact that materials with different
refractive indices will have a profound effect on the pathlength of
rays through the chamber. Figure 4 also demonstrates that the
pathlength through the chamber increases as the angle Θ_c increases.
In other words, the higher the refractive index of the material
filling the chamber, the longer the effective pathlength. The

extreme-ray pathlength through the chamber filled with carbon tetrachloride was calculated to be 27.24 µm as compared to the actual chamber thickness of 25.00 µm assumed in this example. Considering the range of refractive indices which are anticipated for different phases of the sample in the chamber (i.e. 1.33 to 1.60), the pathlength of cell for each phase could vary as much as 2%.

During this investigation a second DIT cell filled with carbon tetrachloride was employed to obtain the reference spectra, under the same conditions used to obtain the sample spectra. In doing so, focusing errors and errors associated with differences in pathlength were minimized such that only those errors associated with the sample cell need be considered. One defect which is known to exist in the cells is the fact that the chamber thickness can be 10% larger at the ends of the cells than at the center. This chamber-thickness variation arises from the manner in which the silica cell parts are polished before the cell is constructed. The DIT-NDX method can be used to measure differences in the cell chamber thickness to a precision of one-hundredth of a micrometer. A plot of the cell thickness as determined by the DIT-NDX method is shown in Figure 5. A profile of the cell pathlength collected by measuring the absorbance of a single phase composition (39.82% (w/w) $C_{10}PO$ in distilled water) over the entire length of the cell is also shown in Figure 5. Both curves exhibit a similar parabolic shape. The ability to precisely calibrate the pathlength of the cell with the DIT-NDX method will be extremely useful for the correction of absorbance values obtained by the DIT-IR technique.

Results and Discussion

For several reasons, some of which were alluded to in the previous section, the preferred reference material for these studies is carbon tetrachloride. First, carbon tetrachloride is transparent in the CH- and OH-stretching fundamental regions and beyond through the entire near-IR portion of the spectrum. Second, its refractive index is 1.46072 at 546.073 nm and 25.0°C while that of silica is 1.46014; using carbon tetrachloride thus serves to eliminate interference fringes. A practical problem which arises from the close refractive index match is that of visually locating the chamber. The sample must, of course, be properly positioned during visual alignment of the cell to obtain a proper spectrum.

A single-beam spectrum from a DIT cell containing carbon tetrachloride is shown in Figure 6. A strong silica band occurs at 3660 cm^{-1} which partially obscures the OH-stretching fundamental region. The cell is opaque below 2100 cm^{-1}. The high-frequency portion of the spectrum is cut off at 7800 cm^{-1}, but this is due to the fact that the interferogram was undersampled by a factor of two. The discontinuity in the data at 5264 cm^{-1} is the result of a low-pass electronic filter in the spectrometer.

Figure 7 shows a series of spectra collected along the length of a DIT cell about a day after the initiation of a phase study of $C_{10}PO$. A spectrum was recorded to either side of each interface within an area of 50 x 160 µm, as well as far from the boundary of the L and X phases and near the middle of the D and E phase bands. Each spectrum was ratioed to a reference spectrum obtained on the cell filled with carbon tetrachloride. The spectra are plotted on

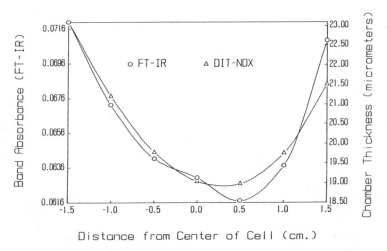

Figure 5. Chamber thickness as determined by the DIT–NDX method and IR absorbance of water at 5174 cm^{-1} as a function of distance from the center of the cell. Pure carbon tetrachloride in a second cell was used as a reference for the IR measurements.

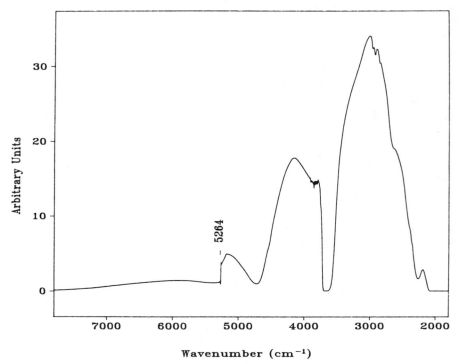

Figure 6. Single-beam spectrum from a DIT cell containing carbon tetrachloride.

the same scale (not offset), in order to illustrate the baseline shifts that occur as spectra are collected along the length of the cell. These shifts are due to refractive-index changes in the sample as a function of composition. At the dilute end of the cell, the refractive index will be that of pure water (1.33) but at the concentrated end it will be significantly higher. It may be as high as 1.5-1.6, depending on the structure, composition, and orientation of the phase. This index variation causes slightly different focusing to exist along the length of the cell, and results in different baseline offsets. We also calculated the contribution of reflective losses to the baseline offset and found them to be insignificant.

The negative peak in the baseline at 4520 cm^{-1} proved to be a convenient reference position. Its origin is presently unknown, but it likely arises from slight differences between the silica DIT cell used for the sample and that used for the carbon tetrachloride reference. Band integration did not work well for quantitation in this study, probably because of uncertainties in the data above 5264 cm^{-1} where the discontinuity due to the electronic filter in the spectrometer occurs.

The OH bend-stretch combination band at 5190 cm^{-1} was used to analyze the compositions. Differences in the absorbance values at 5190 and 4520 cm^{-1} were measured, and the % H_2O estimated assuming that the spectrum with the most intense 5190 cm^{-1} band was that of pure water. This assumption is consistent with the observation that in this spectrum the absorbance at 2930 cm^{-1} (the wavenumber of the most intense surfactant fundamental) is very weak (75 mA). Data were collected to each side of each interface and also near the middle of each phase band. Table I summarizes the estimated composition data, in % $C_{10}PO$. No attempt was made to correct the absorbance values for cell-thickness variations along the length of the cell. Figure 8 shows a scale drawing of the entire 55-mm length of the cell chamber indicating the distances where the interfaces were observed. The arrows along the top of Figure 8 point to key positions on either side of each interface where IR measurements were obtained. The % $C_{10}PO$ calculated from the IR data for each position is indicated.

Table I. Calculated Compositions (% $C_{10}PO$)
at Specific Positions Along the DIT Cell

Phase	% $C_{10}PO$	
	Measured	Phase diagram estimates
Dilute L	(0)	--
Concentrated L	43.1	45
Dilute E	44.2	46
E	51.5	--
Concentrated E	53.8	60
Dilute D	59.0	71
D	77.5	--
Concentrated D	84.5	82
Dilute X	100	100
X	100	100

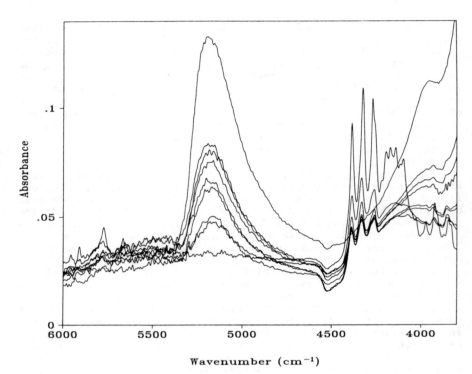

Figure 7. Nine spectra collected along the length of a DIT cell about one day after initiation of a phase study of the decyldimethylphosphine oxide/water system. Pure carbon tetrachloride in a second cell was used as a reference.

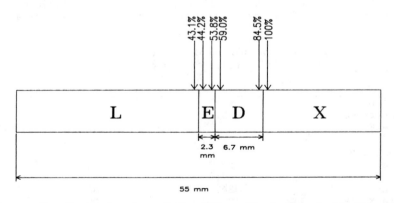

Figure 8. Scale drawing of the DIT cell chamber indicating the distances where the interfaces were observed (bottom). The arrows along the top indicate key positions where IR measurements were made and the % decyldimethylphosphine oxide calculated from the IR data for that position.

The estimated composition data in Table 1 are monotonic along the cell, as must be the case. (In contrast, refractive indices occasionally do not vary in a monotonic fashion because of index inversions at certain interfaces (7).) Further, the estimated compositions of the coexisting phases are in reasonable agreement with those expected from the phase diagram (Figure 1). It must be recognized that the phase study methods used in determining this diagram are capable only of defining the boundary of the liquid phase; it has never been possible to accurately define the boundaries of liquid crystal phases. It was believed that the crystal coexisting with the D phase is the dry crystal (not a crystal hydrate).

The estimated solubility of this compound is in good agreement with the phase diagram. The composition spans of the L–E and D–E miscibility gaps are also reasonable. The DIT–IR experiment predicts a miscibility gap between the L and E phases ranging from 43 to 44%, and a gap between the D and E phases from 54 to 59% $C_{10}PO$. The estimated composition of the D phase coexisting with the dry (100%) crystal is about 85%. The spectroscopic data indicate that the crystal phase is the dry crystal, because no water absorption was observed even for the OH–stretching fundamentals. This is a difficult point to establish during phase studies, and hard evidence that this was true had not previously existed.

In addition to providing data on the compositions of coexisting phases, infrared data is also of potential value in providing structural information about phases. The mid–IR spectral region, particularly the CH–stretching and –bending regions, has been extensively used for this purpose (17–18). The near–IR bands we observe between 4500 and 4000 cm^{-1} are the CH bend–stretch combination bands. It seemed likely that their band positions and intensities could also be sensitive to phase structure.

Figure 9 shows an expanded view of the two spectra to either side of the D–X interface. The top spectrum is that of the dry crystal X, while the bottom spectrum is that of the coexisting D phase. The peak positions of the three largest bands are significantly different in the two phases. Thus, as expected, both this region and the CH–stretching bands hold promise as a means of characterizing phase structure. Such information is useful for characterizing both equilibrium and metastable phases.

These data suggest that the CH bend–stretch combination bands between 4500 and 4000 cm^{-1} might also be used for determining composition. However, composition analysis using this region was not as consistent with the expected data as was analysis using the 5190 cm^{-1} OH–combination band. This is believed to be due to the fact that the OH–stretching fundamental overlaps these CH combination bands and interferes with the analysis, particularly at the dilute end of the cell. In addition, band shifts and intensity changes as a function of phase structure are more likely to be significant for CH bands than for water bands. These breakdowns in Beer's law will compromise quantitative analysis using these bands.

Although many factors will affect the data on a given sample, careful consideration of the sample and characterization of the cell prior to analysis should make the technique analytically viable. In order to investigate the precision of the technique, six spectra were obtained on a 39.82% (w/w) solution of $C_{10}PO$ in a DIT cell at a

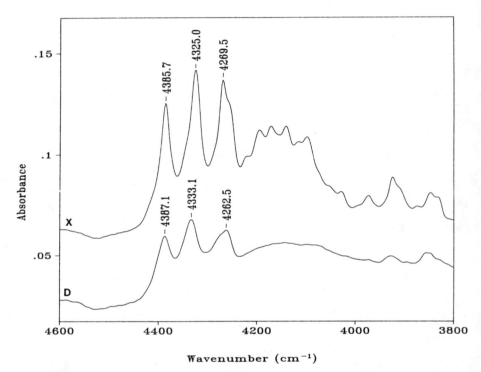

Figure 9. Expanded view of spectra to either side of the D–X interface in the decyldimethylphosphine oxide/water system. The top spectrum is of the dry crystal (X), while the bottom spectrum is of the coexisting lamellar liquid crystal (D) phase.

single position. The water absorbance values at 5174 cm   were found to be 8.3 +/- 0.1 milliabsorbance units, at the 95% confidence limit. The reproducibility of these data are good, considering that the beam splitter and detector employed for this investigation are working beyond their intended wavelength limits.

Future Work

This study demonstrates the feasibility of the DIT-IR phase studies method using a near-IR microscope apparatus. Reasonable results were obtained on the $C_{10}PO$/water model system, even using a mid-IR spectrometer which was not optimized for work in the near-IR region. In future work, a near-infrared FT-IR spectrometer with a CaF_2 beam splitter and a microscope with a small-area InSb detector will be utilized. Precise DIT-NDX measurements of the cell-chamber thickness will also be used to correct all the DIT-IR results for cell pathlength variations along the length of the cell. Such a correction was not performed on the data presented here. Very precise temperature control (+/-0.1° C) will also be added for all future work. With these components and with proper system development, DIT-IR is expected to be an exceedingly useful phase studies method.

Acknowledgment

We thank Richard L. Munyon for preparing the samples used in this work.

Literature Cited

1. Glasstone, S. Textbook of Physical Chemistry; D. Van Nostrand Co., Inc.: New York, 1946.
2. Prigogine, I.; Defay, R. Chemical Thermodynamics; Everett, D. H. Ed.; Longmans, Green & Co., Ltd.: London, 1954.
3. Ekwall, P. Advances in Liquid Crystals; Brown, G. H. Ed.; Academic Press, Inc.: New York, 1975 ; pp 1-142.
4. Madelmont, C.; Perron, R. Colloid and Polymer Sci., 1976, 254, 6581.
5. Kekicheff, P.; Grabielle-Madelmont, C.; Ollivon, M. J. Coll. Int. Sci., 1989, 131, 112.
6. Blackmore, E. S.; Tiddy, G. J. T. J. Chem. Soc., Faraday Trans. 2, 1988, 84, 1115.
7. Laughlin, R. G.; Munyon, R. L. J. Phys. Chem., 1987, 91, 3299.
8. Laughlin, R. G.; Marrer, A. M.; Marcott, C.; Munyon, R. L. J. Mic., 1985, 139, 239.
9. Laughlin, R. G.; Munyon, R. L. J. Phys. Chem., 1990, 94, 2546.
10. Infrared Microspectroscopy: Theory and Applications; Messerschmidt, R. G.; Harthcock, M. A., Eds.; Marcel Dekker: New York, 1988.
11. Yang, P. W.; Mantsch, H. H. J. Coll. Int. Sci., 1986, 113, 218.
12. Herrmann, K. W.; Brushmiller, J. G.; Courchene, W. L. J. Phys. Chem., 1966, 70, 2909.

13. IR-PLAN   Instruction   Manual,   Spectra   Tech.   Inc.,   Stamford,
    Connecticut.
14. Messerschmidt, R. G. In ATSM, STP 949; Roush, P. B., Ed.;
    American Society for Testing and Materials: Philadelphia, PA,
    1987; pp 12-26.
15. Messerschmidt, R. G. In Infrared Microspectroscopy: Theory and
    Applications; Messerschmidt, R. G.; Harthcock, M. A., Eds.;
    Marcel Dekker: New York, 1988, pp 1-19.
16. Katon, J. E.; Sommer, A. J.; Lang, P. L. Appl. Spectrosc. Revs.
    1989, 25(3-4), 173.
17. Mantsch, H. H.; Cameron, D. G.; Umemura, J.; Casal, H. L. J.
    Mol. Struct., 1980, 60, 263.
18. Mantsch, H. H. J. Coll. Int. Sci., 1986, 113, 223.

RECEIVED September 5, 1990

Chapter 6

Micellar Sphere to Rod Transitions

Jeffry G. Weers and David R. Scheuing

Clorox Technical Center, 7200 Johnson Drive, Pleasanton, CA 94588

Micellar growth induced by changes in electrolyte concentration, micelle composition, or temperature, can be monitored by changes in the frequency and shape of CH_2 stretching and deformation bands. Decreases in the *gauche* conformer content of the methylene chains of surfactant molecules with increasing aggregation number are observed. Simultaneous changes in bands due to surfactant headgroups can distinguish differing mechanisms of micelle growth because of their sensitivity to the location and type of counterion species present. The spectroscopic data support the spherocylindrical model of micelle growth, and show that surfactant molecules in the cylindrical portion exhibit methylene chain ordering similar to the *rotator* phase of alkanes.

The utility of FTIR spectroscopy in studies of phase transitions involving micellar surfactant solutions has been increasingly demonstrated in recent years (*1-7*). The familiar concentration-dependent monomer to micelle transition (cmc), the thermally induced hydrated solid to micelle transition (cmt), and the micelle to coagel transition achieved at high pressure (ccp) have all been investigated (Figure 1). These transitions can be monitored by shifts in frequency (≈ 5 cm^{-1}) of the CH_2 stretching bands which appear in the spectra.

Only relatively small shifts of these bands (≈ 1 cm^{-1}) as a function of concentration or temperature are observed in the spectra of micellar surfactant solutions in the absence of the transitions mentioned above. It has therefore been postulated that the methylene chains comprising the tails of surfactants are in a highly disordered "liquid-like" state which does not vary greatly with micelle structure or shape. For example, Mantsch *et al.* reported no differences in the spectra of micellar alkali hexadecylsulfates with temperature or counterion variations (*4*). Cameron *et al.* concluded that the methylene chain packing was the same in spherical sodium laurate and rod-shaped sodium oleate micelles (*3*). An early Raman spectroscopic study of a micelle sphere to rod transition did, however, suggest that there may indeed be changes in methylene chain packing accompanying the shape change (*8*).

As we demonstrate in this chapter, FTIR spectroscopy can be an ideal technique for studying the molecular packing in micellar aggregates. The major

0097–6156/91/0447–0087$10.00/0

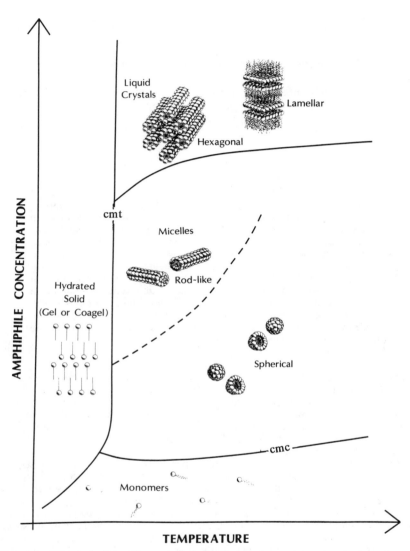

Figure 1: Pictoral phase diagram for a typical ionic surfactant. Micellar phases exist at temperatures above the critical micellization temperature (cmt), and at concentrations above the critical micellization concentration (cmc). A *"pseudophase"* transition from spherical to rodlike micelles may also occur at low temperatures or high surfactant concentrations. Also shown are regions where hydrated solid (gel or coagel) phases and liquid crystals (lamellar or hexagonal) appear (artwork courtesy of Linda Briones).

advantages of FTIR spectroscopy in such studies include the fact that "probe" molecules are not strictly required, and perhaps more importantly, that the time scale of the vibrational modes is quite short. Thus, unlike NMR spectroscopy, FTIR spectroscopy does not suffer from losses of information due to chemical exchange in studies of micelles. If there are two sites available for a solubilizate in a micelle, both will be detectable by FTIR (assuming the existence of an affected vibrational mode, such as a C=O stretch), while the NMR spectra obtained may only indicate an average of the two.

Although a number of infrared bands can be used to establish that a micellar shape change has occurred, it is difficult to determine the actual shape unambiguously from the spectroscopic data alone. We therefore make use of micelle aggregation numbers and solution rheological properties, which depend on micelle size and shape, for correlation with the structural information (packing) provided by the FTIR spectra.

Experimental

Materials. Sodium dodecylsulfate (SDS) and fully deuterated sodium dodecylsulfate (SDS-d_{25}) were obtained from Sigma and Cambridge Isotope Laboratories respectively, and used as received. The cationic surfactants, dodecyltrimethylammonium chloride (DTAC), dodecyltrimethylammonium bromide (DTAB), and didodecyldimethylammonium bromide (DDAB) were purchased from Eastman Kodak, and purified by repeated recrystallization from an ethanol/acetone solvent pair. Even so, a small amount of surface active impurity was observed in surface tension plots for DTAC. The tetradecyldimethylamine oxide ($C_{14}AO$) was a commercial sample (Ammonyx MO) obtained from Stepan (Control No. 533-30027). This sample is primarily $C_{14}AO$, but also contains other chain length components. Sodium chloride (NaCl) was obtained from EM Science and used as received. Water was purified by a three stage Barnstead water purification system.

Methods. Spectra were obtained with a Digilab FTS 15/90 spectrometer equipped with a liquid nitrogen cooled wide band mercury-cadmium-telluride detector. Spectra of the aqueous sample solutions were obtained with the coaddition of at least 256 scans at a nominal resolution of 4 cm^{-1}, using a triangular apodization function. A series of water spectra at the same temperatures were collected each day, and the spectra of the surfactants were obtained by subtracting the appropriate liquid water spectrum. Samples were held in a temperature controlled transmission cell (Harrick TLC-M25) between calcium fluoride windows, at pathlengths of 15 or 25 micrometers set by a spacer. A water/ethylene glycol mixture from a circulating bath (VWR Model 1140) was pumped through the jacket of the sample cell holder. Sample temperatures were monitored with a thermocouple placed in contact with the edge of the windows. Sample equilibration time after each temperature change was 20 minutes. Equilibration was checked by subtracting sequentially acquired spectra from each other and noting an essentially flat difference spectrum. The sample cell was disassembled and cleaned prior to loading of each sample.

Analysis of the shifts in the symmetric CH_2 and CD_2 stretching bands as functions of mixed micelle composition was performed with a peak-finding program (pkprogs) developed by D.J. Moffat at the National Research Council of Canada. This program utilizes a Fourier interpolation/zero slope algorithm which locates peak maxima. For the spectra obtained in this study, which exhibited excellent signal to noise ratios, the use of this or a center of gravity algorithm would be expected to yield similar results (9). The standard software supplied by Digilab for peak location yielded similar trends in the shifts of the CH_2 band with micelle composition.

The same linear baseline corrections were applied to spectra undergoing direct comparisons, before application of the peak finding programs. Typically, baseline points at 3000 and 2800 cm^{-1} for CH_2 stretching bands and 2280 and 2040 cm^{-1} for CD_2 stretching bands were chosen. Cameron *et al.* (9) have demonstrated that frequency measurements with uncertainties of hundredths of a wavenumber or less can be routinely made, independent of instrument resolution. Precision experiments conducted in our laboratory on nine independent SDS solutions in which the cells were demounted and cleaned between experiments, yielded a 95% confidence interval for the frequency of the symmetric CH_2 stretching band of ± .04 cm^{-1}. Similarly, measurement of successive spectra at various spectral resolutions, and on different days gave confidence intervals of ± .01-.02 cm^{-1}. These uncertainties reflect the errors in band frequency determination due to spectrometer drift, temperature control, and any unknown chemical heterogeneity of the samples.

All spectral plots shown were produced with a Digilab 3240 computer. Linear baselines have been subtracted in some cases to facilitate presentation. Some spectra have been interpolated by a factor of 4 for presentation purposes. No noise reduction or smoothing algorithms have been applied to any of the spectra used in these studies.

Measurements of the zero shear viscosity (20 °C) were made with a Bohlin VOR rheometer in the viscometry mode. If a Newtonian region was not found at the lowest measurable shear rates, the samples were characterized with a Bohlin-CS constant stress rheometer, with which it was possible to obtain extremely low shear rates. This approach was especially needed for highly viscous samples exhibiting pseudoplastic behavior on the VOR rheometer. Newtonian regions were found for each sample in this manner, yielding the zero shear viscosities reported.

Spectral Regions Useful for Surfactant Studies

Methylene Tail Modes. Table I summarizes the major infrared active bands due to vibrations of the methylene and methyl groups in the tails of common surfactants. The CH_2 stretching bands near 2920 and 2850 cm^{-1} are generally the strongest bands in the spectra. The frequencies and widths of these bands are sensitive to the *gauche/trans* conformer ratio of the methylene chains, exhibiting a shift from lower frequencies characteristic of highly ordered, *all-trans* conformations, to higher frequencies and increased widths as the number of *gauche* conformers (the "disorder" of the chains) increases. In the case of transitions such as the cmt, or gel to liquid crystal transitions, the frequency shifts reflect the melting of the methylene chains, which is a relatively large change in terms of the *gauche/trans* conformer ratio (3-6). As discussed further below, the changes in average chain conformation accompanying micellar shape changes are more subtle, and hence the frequency shifts of the CH_2 stretching bands are of smaller magnitude. Difference spectra can be used to confirm the small shifts, however, and these yield the same trends in frequency with order to disorder changes as discussed above.

The CH_2 bending or "scissoring" band is known to be extremely sensitive to interchain interactions (3,5). A sharp, narrow singlet is found in the spectra of systems in which the methylene chains are packed in an all parallel arrangement described as a triclinic subcell. The band is split into a doublet when the chains adopt orthorhombic subcell packing, in which the planes of the carbon-carbon bonds of neighboring chains are oriented at about 90 degrees to each other. The third subcell exhibited is the so-called hexagonal type, which is characteristic of the relatively more disordered "rotator" phase of alkanes. The interactions of the chains are considerably smaller in the hexagonal subcell, due to higher concentrations of *gauche* defects which are incorporated in the chains, particularly near their ends.

Table I. Band Assignments of the Methylene Tails

wavenumber (cm⁻¹)	Assignment
2956	ν_{as}, CH_3-R
2920	ν_{as}, CH_2
2870	ν_s, CH_3-R
2850	ν_s, CH_2
1472 (singlet)	δ, CH_2 (triclinic subcell)
1472+1462 (doublet)	δ, CH_2 (orthorhombic subcell)
1468 (singlet)	δ, CH_2 (hexagonal or rotator phase)
1468-1466	δ, CH_2 ("liquid-like" disorder)
1465-1456	δ_{as}, CH_3-R
1420	δ, α-CH_2
1380	δ_s, CH_3-R
1367,1300	w, CH_2 (-*gtg'*-, "kink" defect)
1354	w, CH_2 (-*tggt*-, "double gauche" defect)
1341	w, CH_2 (-*tg*, "end-gauche" defect)

ν = stretching, δ = bending (deformation), w = wagging

The scissoring band of chains packed in the hexagonal subcell is a somewhat broader singlet near 1468 cm⁻¹. Fully disordered liquid-like chains exhibit a considerably broadened and relatively lower intensity scissoring band between 1468 and 1466 cm⁻¹. The scissoring band of methylene groups adjacent to a polar group (α-CH_2) such as an ester or acid carbonyl, or a positively charged nitrogen of a cationic surfactant is often found near 1420 cm⁻¹.

The extensive studies of Snyder *et al.* (*10*) on the infrared spectra of hydrocarbons have also identified several CH_2 wagging bands, which are nearly constant in frequency in the spectra of compounds with various chain lengths, that appear to be related to certain conformations, and are hence known as *"defect"* modes. Changes in the relative intensity of these bands have been used to estimate the conformational disorder of the tails of micellar SDS (*11*), and the changes in packing of solid hydrocarbons as a function of temperature (*12*).

Due to their low inherent intensity and overlap by headgroup modes, the complete set of the CH_2 wagging and the twisting-rocking progressions are generally not useful in studies of surfactants.

The use of deuterated methylene chains can be extremely useful in samples which contain more than one surfactant (mixed micelles), or in mixtures of surfactant plus hydrocarbon solubilizate. The CD_2 stretching frequencies occur approximately at $(2)^{-1/2}$ times the CH_2 stretching band frequencies (i.e. in the 2000-2300 cm^{-1} region), a particularily useful region since it is generally devoid of other interferring absorption bands. Similar to the frequency of the C-H stretching bands, the C-D stretching bands are also sensitive to conformational changes of the methylene tails. Bands near 2194 and 2089 cm^{-1} are assigned to the antisymmetric and symmetric CD_2 stretching bands respectively, while bands near 2212, 2169, and 2070 cm^{-1} are due to R-CD_3 stretching modes.

Headgroup Modes. Table II summarizes the bands due to headgroup vibrations of the surfactants to be discussed in this chapter, i.e. it is not an exhaustive list. As discussed further below, the combination of spectroscopic data from both the headgroups and tails of micellar surfactants of different size and shape can provide information about molecular packing.

Table II. Band Assignments of the Surfactant Headgroups

wavenumber (cm^{-1})	assignment
3040	ν_{as}, CH$_3$ of -N(CH$_3$)$_3$
2985	ν_s, CH$_3$ of -N(CH$_3$)$_3$
1490,1480	δ_{as}, CH$_3$ of -N(CH$_3$)$_3$
1405	δ_s, CH$_3$ of -N(CH$_3$)$_3$
$\approx$1250-1200	ν_{as}, S-O of -SO$_4$
1250-1200	ν_{as}, C-N$\rightarrow$O of amine oxides
1065	ν_s, S-O of -SO$_4$
$\approx$970-950	ν_{as}, C-N-C of amine oxides, quats

Background on Ionic Micelle Structure

Geometric Arguments. Substantial micelle growth beyond the case of a sphere with a radius equaling the length of an extended surfactant molecule corresponds to a micelle shape change. The structure of the micellar aggregate formed is governed by a delicate balance between attractive and repulsive terms of the surface free energy. Simple geometric arguments are effective at predicting the micelle structure which forms. The packing parameter developed by Israelachvili et al. ($P = v/a_o l$) allows the prediction of aggregate structure with three adjustable parameters: a_o, the effective headgroup area, and v and l, the the methylene chain volume and length, respectively (13). When $P \leq 1/3$, spherical micelles are favored, while at $P = 1/2$, infinite rodlike micelles are preferred. At intermediate values of P, a wide range of micelle structures are possible, ranging from globular micelles with oblate shapes to more rodlike prolate shapes. For $P = 1/2\text{-}1$, bilayered aggregates such as vesicles or lamellar liquid crystals are favored. The packing parameter indicates that the formation of elongated micelles or bilayered structures is promoted by lowering the effective headgroup area or by increasing the volume of the methylene chains.

Addition of inorganic electrolytes, such as sodium chloride, to ionic surfactants results in rod micelle formation via reductions in a_o due to enhanced screening between the ionic headgroups (*14*). Similar effects can be achieved with solubilizates that affect packing in the headgroup region. For example, aromatic organic counterions, such as sodium salicylate, bind very strongly to oppositely charged cationic surfactant micelles, and are very effective at promoting rod micelle formation (*15*). Long chain aliphatic alcohols can interact with ionic surfactants through ion-dipole interactions to reduce a_o and promote rod micelle formation (*16*). Similar reductions in a_o can be achieved by mixing surfactants, which have significant interactions (electrostatic, ion-dipole, or even hydrogen bonding) between dissimilar headgroups (*17-19*). Increasing the effective volume of the methylene tails can be achieved in surfactant mixtures by adding increased amounts of a dialkyl surfactant to a monoalkyl surfactant (*20*). Raising the surfactant concentration (*21*), or lowering the temperature (*14*) have also been shown to promote rod micelle formation in certain systems.

Ionic Micelle Structure. A sodium dodecylsulfate (SDS) micelle in water contains roughly sixty SDS molecules (*14*), and is represented by the pictoral diagram in Figure 2. The negatively charged headgroups are located in the *Stern layer*, along with adsorbed or "bound" counterions. Typically only $\approx$ 60-80% of the counterions are bound at any given time. Therefore, to preserve the electrical neutrality of the solution excess "unbound" counterions must still be associated with the micelle. The unbound counterions and any added electrolytes (coions) reside in the so-called *diffuse double layer* (or *Gouy-Chapman layer*). The micelle core contains the surfactant hydrocarbon chains.

Fully saturated alkyl chains are extremely flexible, with a difference in energy between a *trans* C-C bond and *gauche*$^+$ or *gauche*$^-$ states of only $\approx$ 0.8 kT at room temperature. Therefore, the methylene chain of a typical surfactant (e.g. SDS) can exist in a large number of low energy conformations in a micellar aggregate.

Considerable experimental evidence (*1-6,11*) suggests that the methylene chains inside of a spherical micelle are almost as disordered as in the bulk liquid state (i.e. they contain a significant proportion of *gauche* conformers). The FTIR spectra of micellar SDS support this assertion, exhibiting CH_2 stretching and scissoring band frequencies which are comparable to those found in the spectra of liquid hydrocarbons (*1-6,11*). A recent quantitative analysis of the CH_2 defect modes of SDS has shown that the disorder of the methylene tails is similar to that found in liquid tridecane (*11*).

There is little doubt that the surfactant headgroups and counterion species are excluded from the micelle core. There is, however, some debate concerning water penetration into the core. Evidence for water penetration into the micelle core generally comes from spectroscopic experiments employing probe molecules (*22-23*). The probe molecules have been found to lie in a partly hydrophilic environment, and this has been interpreted as indicating water penetration into the core. Recent NMR relaxation (*24*) and neutron scattering (*25*) data provide fairly unequivocal evidence for minimal water penetration into the micelle core, however.

A theoretical model has been developed by Gruen (*26*) which helps to harmonize the apparent experimental discontinuities. The model assumes the existence of a hydrocarbon core completely devoid of headgroups, counterions, and most importantly, water. The methylene chains are then allowed to exist in a number of conformations (*trans, gauche*$^+$, and *gauche*$^-$ states for each C-C bond), and are, on average, constrained to pack at liquid hydrocarbon density. Each chain conformation is assigned its appropriate Boltzmann factor, and the methylene chains are averaged over all conformation space. Surprisingly, the theory predicts that all chain segments spend an appreciable proportion of time near the micelle surface, and although the micelle core is completely devoid of water, each segment of the methylene chain samples a partly hydrophilic environment. On average, only one or two molecules are able to exist in an *all-trans* conformation extending into the micelle center for spherical micelles. There is no room for more fully extended molecules if the packing density is to be uniform throughout the micelle. The methylene chains pictured in Figure 2 are drawn to illustrate this concept. Experiments employing a probe molecule could indeed detect a hydrophilic environment in the absence of extensive water penetration into the micelle core. It is possible that these probe experiments provide more information about methylene chain dynamics than water penetration.

Several other groups have also modeled chain organization in amphiphilic structures. Dill and Flory (27) constructed a mean field lattice model in which the methylene chains are placed on a three dimensional lattice and probabilities of various conformations are calculated. Szleifer, Ben-Shaul, and Gelbart (28) have calculated molecular and thermodynamic properties for methylene chains in various aggregate shapes. Woods *et al.* (29) have used molecular dynamics to show that a spherical micelle containing a dodecyl methylene chain has approximately 28% *gauche* conformers. Flory's rotational isomeric model (30) predicts that a liquid hydrocarbon of similar chain length would have about 36% *gauche* conformers. This difference, as well as other calculations suggest that the methylene chains in a micelle interior may be slightly more ordered that in the corresponding liquid alkane. Although there is a limited amount of direct experimental data concerning average methylene chain conformations in small spherical micelles, little or no direct information exists for the corresponding rodlike micelle. The work detailed in this chapter attempts to bridge this gap.

Transition Moment Vectors for the Headgroup Modes. Also pictured in Figure 2 are the transition moment vectors for the S-O stretching modes in SDS. The transition moment vector for the symmetric S-O stretching mode points approximately normal to the micelle surface, and are extremely sensitive to changes in the location and distance of Na^+ counterions in the Stern and diffuse double layers. The transition moment vectors for the asymmetric S-O stretching vibrations, on the other hand, point nearly parallel to the micelle surface, i.e. in the direction of neighboring surfactant molecules. Changes in the asymmetric mode can, therefore, be linked with changes in the surfactant-surfactant interactions. These differences in directionality of the transition moment vectors can be used to great advantage in detailing the factors which give rise to micellar structural changes, as detailed below.

Micelle Sphere to Rod Transitions. There are at least two competing effects responsible for determining a transition of micelle shape from spheres to rods. First, simple geometric arguments predict that the area per headgroup will decrease upon going from a sphere to a rod-like micelle. For example, in SDS the area per surfactant headgroup on the surface of the minimum spherical micelle is equal to 62 $Å^2$/molecule, assuming that the hydrocarbon chain is fully extended and packs at liquid hydrocarbon density throughout. This compares with the value of 41 $Å^2$/molecule for a cylinder under comparable conditions (assuming equal volumes). This decrease in a_o results in a higher surface charge density for a cylindrical structure relative to the sphere, an effect which opposes the transition from spheres to rods by making a positive electrostatic contribution to the free energy of rod micellization. On the other hand, the closer packing of the headgroups in a cylindrical structure is thought to provide a driving force for rod formation by reducing the residual hydrocarbon/water contacts (hydrophobic effect). Although several models (*31-33*) have broken down the free energy of rod micelle formation into contributions from electrostatic and hydrophobic terms, it is difficult to quantify each term, due to difficulties in modeling either interaction.

Debye and Anacker (*34*) have developed a model which treats the rod micelles as spherocylinders, in which two hemispherical endcaps approximate a spherical micelle (Figure 3). The additional molecules (beyond the number required for a spherical structure) are packed into a cylinder separating the endcaps. Leibner and Jacobus (*35*) have shown that, from geometric considerations, the spherocylinder is the most acceptable structure because it has the highest *average* area per headgroup of any of the nonspherical structures. Mazer and coworkers (*14,31*) have shown that a spherocylindrical model can quantitatively account for changes in aggregation numbers and polydispersity of micelles as functions of temperature and electrolyte level in SDS/NaCl mixtures.

In the spherocylindrical model, two distinct environments for SDS molecules are predicted. The hemispherical endcaps would approximate the environment found in spherical SDS micelles in the absence of electrolyte, while molecules in the second environment (cylindrical portion) would be characterized by closer packed headgroups, and as such might be expected to be "more ordered" than the molecules in the hemispherical endcaps. The model of Gruen (*26*) suggests, however, that there is no energy cost, as far as the methylene tails are concerned, in forming rod-like micelles (i.e. the tails are still able to pack at liquid hydrocarbon density throughout).

One of our goals in applying FTIR spectroscopy to the study of micelle shape changes is to obtain data which addresses simultaneously methylene chain ordering (conformation), and headgroup spacing, for a comparison with these models.

Results and Discussion

Electrolyte Effects. Micelle sphere to rod transitions can be induced in SDS by the addition of electrolytes such as NaCl, which are thought to shrink the diffuse double layer thickness, thereby screening the electrostatic repulsions between the charged sulfate headgroups. This is not meant to imply that the reduction in electrostatic repulsions is the driving force for the sphere to rod transition. It is not, since the overall electrostatic contribution is still highly unfavorable. Rather, the addition of electrolyte is able to reduce the electrostatic contribution just enough so that the hydrophobic effect can take over to drive the sphere to rod transition (*31*). In terms of the micelle aggregation number (N), a small increase from 60 to 135 is observed at NaCl levels up to 0.4 M, and constitutes a transition from spherical to slightly aspherical (globular) micelles (*36*). At approximately 0.45 M, a large increase in N corresponding to the major transition from globular to rodlike micelles is observed.

Methylene Tail Modes. Figure 4 presents representative spectra in the C-H stretching region of 70 mM SDS as a function of added NaCl (T=25°C). Also plotted are several difference spectra, obtained by subtraction of the spectrum of SDS in water from the spectra of the SDS samples containing added electrolyte, and a spectrum of a SDS coagel phase (T=6°C).

The inset shows that the frequency of the v_s CH$_2$ band exhibits a decrease with increasing salt level. Although the frequency shift of the band is smaller than that found in the case of other micellar transitions, such as the monomer to micelle (*1-2*), or the coagel to micelle transition (*3-6*), it should be recalled that this pseudophase transition (sphere to rod) represents a smaller structural change than in the case of transitions involving two distinct phases, such as micelles and coagels. Frequency shifts of the v_s CH$_2$ band of 1 cm^{-1} have been correlated with large structural changes in the case of phospholipids (*37*). The transition of egg yolk

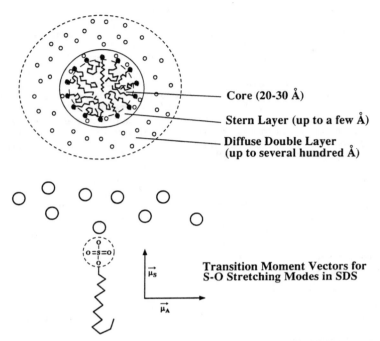

Figure 2: Conventional representation of micelles formed by an ionic surfactant, such as sodium dodecyl sulfate. The inner core region consists of the methylene tails of the surfactants. The Stern layer consists of surfactant headgroups and bound counterion species. The diffuse double layer consists of unbound counterions and coions which preserve the electrical neutrality of the overall solution. Also pictured are the transition moment vectors for the S-O stretching modes of sodium dodecyl sulfate.

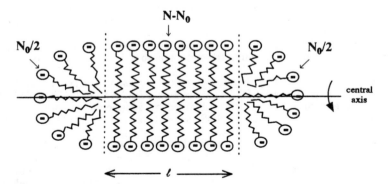

Figure 3: The structure of the prolate spherocylindrical micelle having an aggregation number N. There are $N_0/2$ molecules in each hemispherical endcap, and $N-N_0$ molecules in the cylindrical portion.

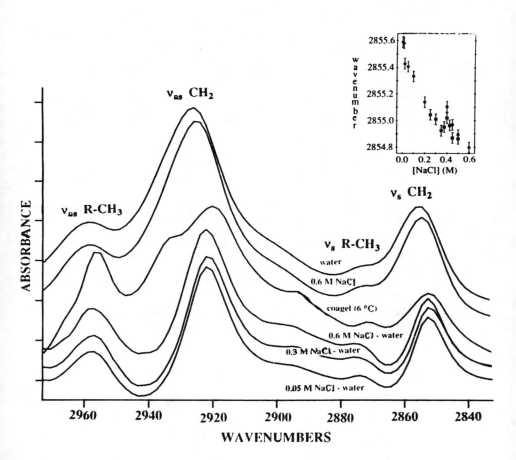

Figure 4:   Plot of the C-H stretching region of 70 mM SDS solutions under various conditions (T = 25°C, except where noted). From top: 0 M NaCl and 0.6 M NaCl micelle solutions, 0 M NaCl (T = 6°C, coagel), and three difference spectra 0.6-0 M, 0.4-0 M, .05-0 M. The inset shows a plot of ν_s CH$_2$ as a function of added NaCl.

phosphatidylethanolamine from the liquid crystalline phase to the inverted
hexagonal phase yielded a band shift of 1 cm^{-1} (37). The frequency precision of
modern FTIR spectrometers, as discussed by Cameron et al. (9), and in our studies,
(illustrated by the error bars in the frequency plot at inset of Figure 4) makes the
determination of such shifts readily detectable.

The difference spectra provide additional information. The 0.8 cm^{-1} decrease
in frequency of the v_s CH$_2$ band upon the addition of 0.6 M NaCl results from the
relative increase in intensity of a highly shifted CH$_2$ band, which appears in the
difference spectrum near 2852 cm^{-1}. This relatively low frequency is comparable to
that found in the spectrum of the coagel phase of SDS in electrolyte-free water at 6
°C (2852.5 cm^{-1}). In the latter, the methylene chains are present in the more ordered,
all-trans conformation.

Difference spectra are obtained by adjustment of a scaling factor, X, applied to
a reference spectrum, to an optimum value in which bands common to both sample
and reference are nulled out. The difference spectrum is then obtained as described
by equation 1.

$$D = S - (X)(R) \qquad\qquad (1)$$

where D is the difference spectrum, S is the sample spectrum (SDS in electrolyte),
and R is the reference spectrum (SDS in water). We refer here to spectra which have
already had the spectrum of liquid water subtracted in the same manner.

The effect of the subtraction technique on the appearance of the difference
spectra can be investigated by plotting the frequency of significant bands in the
difference spectra (such as the v_s CH$_2$ band mentioned above and the v_s S-O band
discussed further below) as a function of the subtraction scalar, X. Table III shows

Table III. Effect of the sub scalar on difference spectra frequency in 70 mM SDS + 0.6 M NaCl minus 70 mM SDS in water.

Sub Scalar	v_S CH$_2$ (cm^{-1})	v_S S-O (cm^{-1})
0.80	2853.50	1065.79
0.90	2853.02	1065.76
1.00	2852.33	1065.42
1.02	2852.23	1065.42
1.10	2851.93	1065.42
1.112	2851.93	1068.77
1.115	2851.00	1068.77
1.116	2851.00	1068.77

that the frequency of the v_s CH$_2$ band changes rapidly for values of X which are far
from the optimum. In contrast, over a narrow range of X, the frequency of the band
in the difference spectrum is not a strong function of X, suggesting that the optimum
value has been selected. In the case of the difference spectrum obtained from SDS in
0.6 M NaCl, the v_s CH$_2$ and v_s S-O band frequencies change little over the range
1.0 < X < 1.10. For greater values of X, sharp frequency shifts of both bands are
noted, and the difference spectra clearly indicate "oversubtraction" by exhibiting a
mixture of positive- and negative- going bands. We cannot attribute a quantitative

significance to the value of X necessary for a "successful" subtraction in these samples because of the disassembly and cleaning of the Harrick transmission cell performed for each sample. However, since the same windows and spacer thickness were used for these samples, which were all of the same concentration and temperature, it is gratifying that the optimum value of X was near 1.0 in all cases, as is expected for the subtle changes in micelle structure under study.

The difference spectra in Figure 4 indicate that addition of NaCl induces frequency decreases for both the v_s and v_{as} CH$_2$ bands. Two factors can contribute to this shift. For SDS molecules incorporated in the cylindrical portion of a rodlike micelle, a partial "straightening" of the methylene chains comprising the tails is required, as discussed above. The decrease in the *gauche/trans* conformer ratio of the chains upon such a change will produce a frequency decrease. In addition, as FTIR spectroscopic studies of the monomer to micelle transition of alkanoates has shown (*1-2*), the decrease in contact of the methylene chains with water upon micelle formation results in a frequency decrease of v_s and v_{as} CH$_2$ due to a refractive index effect. The formation of rodlike micelles upon addition of salt is expected to reduce the contact of water with the micellar core/water interface, and thus the frequency shifts observed might be attributed to changes in water/SDS tail contact. Our examination of changes in other vibrational modes (CH$_2$ scissoring and "defect" modes) in the spectra of rodlike SDS micelles and other systems described below suggest that partial methylene chain "ordering" or straightening is primarily responsible for the spectroscopic changes.

The spectrum of the SDS coagel included in Figure 4 is quite distinct from those of the rodlike and spherical micelles. The higher ordering of the methylene chains, compared to that in the case of micelles, results in lower frequencies for the CH$_2$ bands. The v_{as} CH$_3$ band near 2960 cm^{-1} is relatively more intense and narrower than in the micelle spectra, due to the restricted rotation of the terminal methyl groups of SDS in the coagel. The intense shoulder near 2935 cm^{-1} can be assigned to the v_s CH$_2$ of the methylene group adjacent to the sulfate group (α-CH$_2$), by analogy with the spectra of alkanoate coagels (*2*). A comparison of the SDS coagel spectrum with the difference spectra thus confirms that the latter are not simply due to the partial precipitation of SDS upon addition of up to 0.6 M NaCl. Spectroscopic and visual detection of the coagel of SDS was observed, however, at 25 C, in the presence of 0.8 M NaCl.

By applying a Fourier self-deconvolution to the micelle and coagel spectra, the change in frequency of the v_s CH$_2$ band which accompanies the sphere to rod transition upon the addition of 0.6 M NaCl can be confirmed. Figure 5 shows deconvolved spectra (obtained using a band width of 15 cm^{-1}, triangular apodization function and a K factor of 2.0) and a difference spectrum obtained from those of SDS in 0.6 M NaCl and SDS in water. The v_s CH$_2$ band is again found to shift 0.8 cm^{-1} toward lower frequency (from 2855.16 cm^{-1} to 2854.36 cm^{-1}) upon addition of salt. The difference spectrum obtained from the deconvolved spectra (with the same X factor as for Figure 4) shows the appearance of a CH$_2$ band at 2852.6 cm^{-1}, to be compared with that of the band in the coagel spectrum at 2852.1 cm^{-1}. The CH$_3$ bands, which are narrowed somewhat by the deconvolution, shift only slightly upon the addition of salt. However, these bands do appear in the difference spectrum, shifted toward somewhat lower frequency. In the deconvolved coagel spectrum, the v_{as} CH$_3$ band is split into a narrower doublet at 2966.5 and 2956.0 cm^{-1}, indicating considerably more hindered rotational motion of the terminal methyl groups (*38*), compared to that possible in the micelles. The appearance of the CH$_3$ bands in the difference spectrum is due to a slight increase in intensity relative to the CH$_2$ bands, and a shift toward lower frequency, which can be attributed to a decrease in the rotational motions of the terminal methyl groups, or a slight increase in the ordering of the chain ends, which is however, not as large as is found in the much more solid-like coagel state. The deconvolved spectrum of the SDS coagel is again quite distinct from those of the micelles and the difference spectrum, resembling instead the deconvolved spectrum of an alkanoate coagel (1-2).

An examination of the changes in the δ_s CH_2 band which accompany the sphere to rod transition induced by addition of salt also indicates increased ordering of the methylene chains in the rodlike micelles. As the NaCl concentration increases, the band shifts toward slightly higher frequency, reaching 1467.9 cm^{-1} in the case of 0.6 M NaCl. This relatively high frequency suggests ordering of the methylene chains approaching that of a hexagonal subcell, i.e. an extended all-trans conformation which contains a few gauche defects, mostly near the chain ends. The increase in the intensity of the δ_s CH_2 band relative to the δ_{as} CH_3 band (shoulder near 1460 cm^{-1}), and the decrease in width at three-quarters height of the δ_s CH_2 band (10.73, 9.13, 8.13 cm^{-1} for 0.0, 0.3, 0.6 M NaCl concentrations respectively) also are quite consistent with increased methylene chain straightening accompanying chain crowding as the number of SDS molecules occupying the cylindrical portion of the rodlike micelle increases with aggregation number.

A comparison of the very weak CH_2 wagging "defect" modes in the spectrum of SDS in water and in 0.6 M NaCl also suggests significant changes in the ordering of the SDS tails (Figure 6). The band near 1354 cm^{-1}, which indicates double gauche defects ("bend" in the chains) is more intense relative to the δ_s CH_3 ("umbrella" deformation) in the spectrum of SDS in water only. The relative intensity of the shoulder near 1367 cm^{-1} ("kink" band) is about the same in both spectra, but the 1341 cm^{-1} band ("end gauche") is found to be enhanced in the spectrum of SDS in 0.6 M NaCl. This comparison provides more evidence for a population of SDS molecules packed in manner which is almost as ordered as a true hexagonal subcell in the cylindrical portion of a rodlike micelle, because, from studies of hydrocarbon "rotator" phases, "kink" and "end gauche" defects are known to be more readily accommodated in ordered arrays of hydrocarbon chains than gauche-gauche defects, which are highly disruptive to lateral chain order (10).

All of the spectroscopic evidence from both the C-H stretching and deformation regions of the spectra of rodlike SDS micelles formed by the addition of salt indicates that a second "type" of environment for the tails of SDS appears, accompanying the sphere to rod transition. The changes in the bands due to both CH_2 and CH_3 groups can be interpreted as indicating a decrease in the gauche/trans conformer ratio, i.e., a partial ordering of the SDS tails, in the rodlike micelles, which contain this second environment. The gradual increase in relative intensity of the shifted CH_2 band with increasing salt level, and the presence of the shifted band in the spectra of SDS solutions of even low salt concentration is consistent with the changes in aggregation number. The appearance of a second environment for the SDS tails upon the addition of salt is consistent with Debye and Anacker's spherocylinder model discussed above (34). The model of Gruen, on the other hand, suggests that there is negligible energy cost, as far as the tails are concerned, in forming rodlike micelles, i.e., the methylene chains still pack at nearly liquid hydrocarbon density in all micelles (26). The source of the discrepancy in these two models lies in the assumption by Gruen of constant headgroup interactions during micelle shape changes. As has been shown in the case of phospholipid bilayers, changes in headgroup packing may have profound effects on the packing of the methylene chains (39). We now turn to the effect of salt concentration on the headgroup bands of SDS, in an effort to distinguish between various models of rodlike micelle formation.

Headgroup Modes. Figure 7 presents fingerprint spectra in the S-O stretching band region. It is interesting to note that there are not large differences in the band shape and frequencies as a function of electrolyte level. This is consistent with the assertions made by Mantsch (4) who found no differences in the S-O stretching modes in micelles with various counterion species. There are, however,

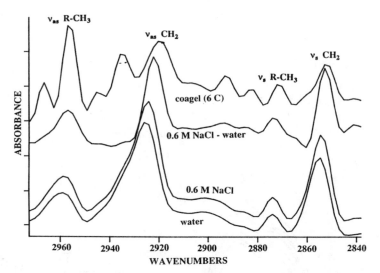

Figure 5:   Deconvolved spectra in the C-H stretching region for 70 mM SDS solutions under various conditions. The bandwidth is 15 cm^{-1} with triangular apodization and a K factor of two.

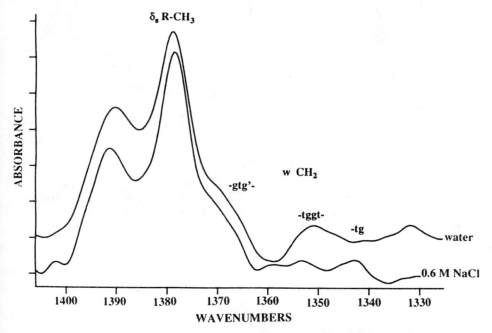

Figure 6:   CH$_2$ defect modes for 70 mM SDS in 0.0 and 0.6 M NaCl at 25°C.

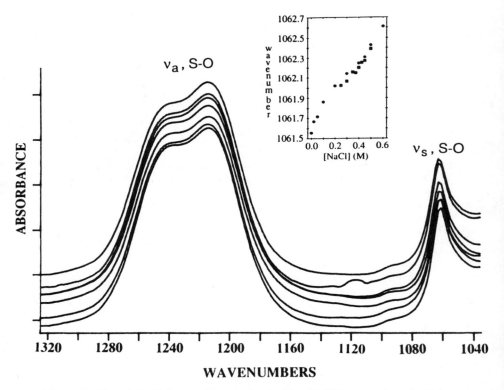

Figure 7: Plot of the S-O stretching region in 70 mM SDS at varying electrolyte levels (T = 25°C). The concentration of added NaCl from top to bottom spectra are 0.6, 0.5, 0.45, 0.2, 0.1, 0.05, 0.0 M. The inset shows a plot of ν_s S-O as a function of added NaCl.

small differences which can be explored in more detail. The symmetric S-O band frequencies vs. [NaCl] are plotted in the inset of Figure 7.

Counterion binding is not a well defined quantity, with various experimental techniques weighing the ion distribution slightly differently. Thermodynamic methods (e.g. ion activities or osmotic coefficients) monitor the free counterion concentration, transport methods (e.g. ion self diffusion or conductivity) the counterions diffusing with the micelle, and spectroscopic methods (e.g. NMR) the counterions in close contact with the micelle surface. Measurement of the effect of Na^+ counterions on the symmetric S-O stretching modes would also be expected to be highly dependent on the distance of the counterion from the micelle surface (similar to the NMR method).

The addition of NaCl to SDS micelles does not alter the counterion binding within the Stern layer appreciably (although a small increase in counterion binding is expected at the change in aggregate shape from a sphere to a rod) (40). Thus, the increase in the frequency of the symmetric S-O stretching band must be related to changes in the distribution of counterions in the diffuse double layer, or possibly to changes in the hydration of the headgroups. Each of these factors will now be examined to ascertain their role in the shift of the symmetric S-O stretching mode.

Addition of NaCl to SDS micelles is thought to reduce the electrostatic repulsions between the charged headgroups by compressing the diffuse double layer, thereby bringing more counterions in close contact with the micelle surface. The effect is fairly dramatic. For example, the Debye length, which is proportional to the diffuse double layer thickness, decreases from 33.8 Å in the absence of electrolyte to 5.54 Å in 0.3 M NaCl, and to 3.5 Å in 0.8 M NaCl, which implies that nearly all of the counterions are in close proximity to the micelle surface at high salt concentrations (14,32). Assuming that the counterion binding is nearly independent of NaCl concentration, the decrease in diffuse double layer thickness would be expected to affect the symmetric S-O stretching frequency to a greater extent, since the transition moment vector for the symmetric mode points normal to the micelle surface towards the diffuse double layer (see Figure 2). The increase in frequency is consistent with an increase in counterion interactions (1,2).

The driving force for formation of rod shaped SDS micelles is the elimination of water from the micellar core/water interface (31). The reduction in average headgroup area reflects the removal of water molecules between the SDS headgroups, and should affect the bands due to the asymmetric S-O stretching vibrations, as indicated in the discussion of the transition moment vectors above.

The asymmetric S-O bandshape in the spectra of micellar SDS is much broader and more complex than that of the symmetric mode because it consists of several overlapping components. A description of the changes in the asymmetric band with a single frequency shift is not as complete as in the case of the symmetric mode. In order to examine the changes in bandshape of the asymmetric S-O stretching band in more detail, the spectrum of spherical SDS micelles (in water) is subtracted from the spectra of nonspherical SDS micelles in electrolyte (Figure 8). In the subtraction an attempt to null out the symmetric and asymmetric S-O stretching bands was made by the appropriate scaling of the spectrum of pure SDS micelles. The difference spectra show significant changes in both the symmetric and asymmetric modes. Two distinct bands of nearly equal intensity are observed at ≈ 1266 and 1216 cm^{-1}, which increase in absorbance with salt concentration. The exact number and frequency of the asymmetric S-O stretching bands depends upon the counterion type and crystal structure in solid alkyl sulfates (4). Solid SDS, for example, exhibits an asymmetric S-O band near 1263 cm^{-1} accompanied by several minor shoulders, and another major band near 1205 cm^{-1}. At the critical micellization temperature these bands broaden and move towards each other to yield the broad band near 1214 cm^{-1} discussed above. The two components obtained in the difference spectra of Figure 8 resemble the major components in the spectrum of solid SDS, suggesting an increasing crowding (i.e. more solid-like character) for the headgroups in the SDS+salt samples, which reflects the decrease in headgroup areas discussed above. The effect of NaCl on some population of SDS headgroups can also be likened to the effect of Ca^{2+} ions on the asymmetric P-O bands of the phosphate groups of fully hydrated phosphatidylcholine bilayers (41). The "local site" symmetry lowering of the phosphate groups by complexation with Ca^{2+} caused the appearance of P-O multiplets resembling the highly ordered, solid state spectra.

The two distinct types of SDS headgroup packing indicated by the difference spectra are consistent with the spherocylindrical model of rod micelle formation discussed above. The increase in absorbance of the difference bands with increasing salt concentration indicates a continual increase in the number of relatively ordered SDS molecules packing into the cylindrical portion of the micelle, with the subtraction procedure simply eliminating the spectral contributions from the hemispherical endcaps.

It is interesting that both the symmetric and asymmetric S-O stretching modes are affected by electrolyte addition. Changes in the symmetric mode occur because of large changes in the Na^+ concentration in the diffuse double layer. The changes in the asymmetric mode, which are expected to be very sensitive to surfactant-surfactant interactions, also show large deviations, presumably because the electrolyte effectively screens the ionic headgroup repulsions, thereby excluding water molecules and allowing the closer approach of the sulfate headgroups. The increased crowding of the sulfate headgroups causes a local site symmetry lowering which leads to increased splitting of the asymmetric modes, i.e. a more solid-like character of the spectra. The methylene tails packed into the cylindrical portion of the spherocylinder also respond by becoming more ordered, i.e. approaching packing in a hexagonal subcell. The methylene chains in the cylindrical portion of the spherocylinder are more ordered than those in the hemispherical endcaps, which have order similar to that of liquid tridecane. The CH_2 stretching and scissoring mode frequencies imply that the ordering of the chains in the cylindrical portion of the micelle is comparable to that of a rotator phase. The hexagonal subcell can accomodate a number of gauche defect structures, most of which will occur near the terminal methyl groups (as evidenced by the defect modes). Thus, although the chains in the cylindrical portion are not *all-trans*, they are much more ordered than the methylene chains in the endcap regions.

Mixed Micelle Effects. Another way to vary the curvature (i.e. induce rod micelle formation) of a micelle is by mixing surfactants with dissimilar headgroups. Practical interest in mixed surfactant systems arises because almost all applications involving surfactants make use of surfactant mixtures (42-43). This is due to the fact that commercial sources of surfactants are generally composed of isomeric mixtures, and mixtures of dissimilar surfactants often show synergistic performance enhancements. In more fundamental studies (44-46), the current literature has focused on theoretical modeling of mixed micelle formation in aqueous solutions. The regular solution theory approximation first applied by Rubingh is the most frequently used model (45-46). The theory contains one adjustable parameter, β, related to the degree of interaction between the two surfactants. If $\beta = 0$, the mixed micelles obey ideal solution theory, and the surfactant-surfactant interactions are constant. Large negative values of β are found for strongly interacting mixtures of dissimilar surfactants (e.g. anionic/cationic surfactant mixtures), while positive values of β due to repulsive interactions between the surfactant tails have been observed in fluorocarbon/hydrocarbon surfactant mixtures (44).

Rod micelle formation is promoted in mixtures of surfactants when there is either a large reduction in the headgroup areas, or an increase in the methylene tail volume, and thus, in terms of the surfactant-surfactant interaction parameters, is favored by large negative values of β. The forces which hold micelles together are not strong covalent bonds, but rather electrostatic forces, hydrogen bonding, and ion-dipole interactions. In the sections which follow, we will examine how these various interactive forces affect surfactant packing in mixed micelles, thereby promoting rod micelle formation.

Variation in a_o: electrostatic interactions. The largest degree of surfactant-surfactant interactions is found in mixtures of anionic and cationic surfactants, where the attractive electostatic interaction between the oppositely charged surfactant headgroups leads to significant nonidealities. In mixtures of SDS with the cationic surfactant dodecyltrimethylammonium chloride (DTAC), β is equal to ≈ -25. What this means in terms of aggregation properties is that a) the mixture cmc is about two orders of magnitude less than either pure surfactant cmc, and b) there is a steep increase in the micelle aggregation number as equimolar ratios of the two surfactants is approached. A recent study by Scheuing and Weers (*47*) has examined the structural aspects of the DTAC/SDS mixed micelles using FTIR spectroscopy. The results are discussed below.

Methylene Tail Modes. Figure 9 shows a plot of the aggregation numbers and the frequency of the symmetric CH_2 stretching band for 0.3 M DTAC/SDS mixtures (25 °C). The CH_2 stretching bands (symmetric and asymmetric) of DTAC and SDS are highly overlapped and, therefore yield a "composite" CH_2 band with contributions from both surfactant species. The frequency of v_s CH_2 decreases as the ratios of the two surfactant components approach equimolar. These shifts are interpreted, as before, as indicating a smaller number of *gauche* conformers in the tails of the mixed micelles compared to the case of the pure component micelles. This decrease in *gauche* conformer content coincides with the sharp increases in the micelle aggregation numbers measured by Malliaris *et al.* (*19*). It is also interesting to note the correlation between v_s CH_2 and N for the SDS-rich compositions, where the larger increase in aggregation numbers is correlated with the larger decrease in composite CH_2 stretching frequencies. The total shift in the composite v_s CH_2 is 1.3 cm^{-1}, which is similar to that observed in SDS+salt mixtures. The 1.3 cm^{-1} shift in v_s CH_2 is due to an increase in absorbance of a component at 2852.7 cm^{-1}. The presence of this highly ordered component of the CH_2 band in the difference spectra implies that the shape change induced by mixing surfactants also conforms to the spherocylindrical model. Experiments in which the methylene tail of the SDS is selectively deuterated showed that each methylene tail (DTAC and SDS-d_{25}) are independently ordered as equimolar ratios are approached. This justifies the use of composite bands to ascertain changes in mixed micelle structure.

Headgroup Modes. The changes in the headgroup region of the DTAC/SDS mixed micelles as a function of composition can also be monitored spectroscopically with the use of difference spectra (Figure 10). Small changes in v_{as} S-O bandshape which are observed in the original SDS-rich compositions are related through the difference spectra to increases in absorbance of a highly split band pair with components at ≈ 1255 and 1206 cm^{-1}. This band pair resembles that found in solid SDS, and in the difference spectra for SDS+NaCl. Unlike the case of SDS+NaCl, where both v_s S-O and v_{as} S-O are highly perturbed and hence appear in the difference spectra, only v_{as} S-O is perturbed to a large extent in the DTAC/SDS mixtures. The small frequency shift observed in the original v_s S-O band appears as a shoulder in the difference spectra, and is attributed to decreases in counterion binding observed previously for DTAC/SDS by Filipović-Vinceković and Škrtić (*48*). The transition moment vector of v_{as} S-O points between surfactant molecules, and is therefore extremely sensitive to surfactant-surfactant interactions. The enhanced perturbation of v_{as} S-O indicates that the electrostatic interactions between the dissimilar headgroups reduces a_o sufficiently to "drive" rod micelle formation in mixtures of DTAC and SDS.

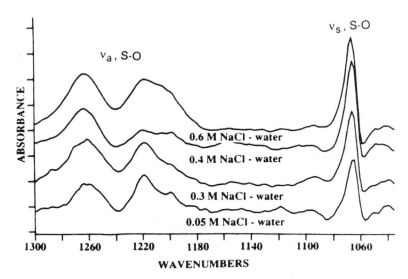

Figure 8: Difference spectra in the S-O stretching region for 70 mM SDS/NaCl mixtures (T = 25°C).

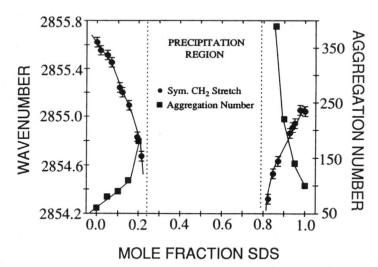

Figure 9: Plot of the frequency dependence of the composite symmetric CH_2 stretching band and micelle aggregation numbers (19) vs. mixed micelle composition in 0.3 M DTAC/SDS mixed micelles (T = 23°C). Reprinted from ref. 47. Copyright 1990 American Chemical Society.

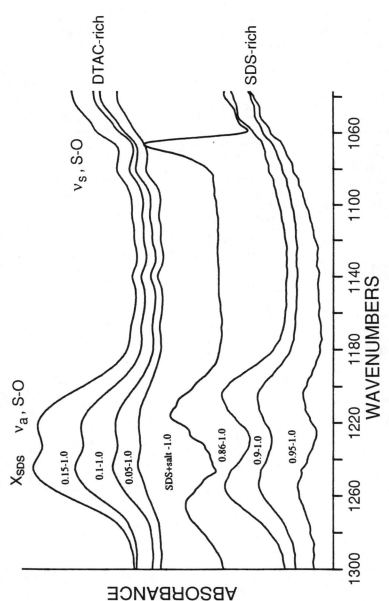

Figure 10: Difference spectra in the S-O stretching region obtained by subtraction of the spectrum of SDS micelles in water (hemispherical endcap environments) from the mixed micelles of 0.3 M DTAC/SDS ($\bar{T}$ = 23°C). Also pictured is a difference spectrum from the SDS/NaCl study. Reprinted from ref. 47. Copyright 1990 American Chemical Society.

For the DTAC-rich compositions, the difference spectra closely resemble the original spectra, implying that nearly all of the SDS molecules are highly perturbed. At these compositions v_{as} S-O is composed of two bands of nearly equal intensity, quite distinct in width and frequency from those of SDS micelles in water. These changes are due to the strong electrostatic attractions between the dissimilar surfactants which leads to increased splitting for the v_{as} S-O modes. The v_s S-O modes also exhibit small decreases in frequency due to an overall change in the counterion species from Na^+ to DTA^+. Therefore, we are able from the difference spectra alone to distinguish the mechanisms for rod growth in the headgroup region, whereby in SDS/NaCl mixtures the decrease in diffuse double layer thickness leads to large differences in both the v_s S-O and v_{as} S-O modes, while in SDS/DTAC mixed micelles the headgroup attraction leads to preferential changes in v_{as} S-O.

A series of bands assigned to the deformation of the methyl groups in the trimethylammonium headgroups of DTAC are also found in the fingerprint region. These bands, along with the CH_2 deformation modes, are shown in Figure 11. The δ_{as} CH_3-N^+ band pair near 1490 and 1480 cm^{-1} found in aqueous solutions of cationic surfactants are known to be sensitive to the disorder and packing of the cationic headgroups. The addition of SDS to DTAC causes slight increases in frequency (towards the frequencies found in solid DTAC) indicating increases in "ordering" of the trimethylammonium headgroups. The difference spectrum shows that the small frequency shifts in the original spectra are due to the increased intensity of bands near 1495 and 1485 cm^{-1} respectively. This change suggests that the addition of small amounts of SDS to DTAC causes a significant change in the electrostatic environment of a population of DTAC headgroups, presumably those which pack into the cylindrical portion of the spherocylinder. The shift towards higher frequency of the band near 1378 cm^{-1} (δ R-CH_3), the development of a shoulder near 1405 cm^{-1}, and the appearance of a band near 1427 cm^{-1} all suggest an increase in ordering of the DTAC headgroups, as well, since these shifts cause the spectra to more closely resemble solid DTAC.

The slight increases in relative intensity and shift towards higher frequency of the CH_2 scissoring and α-CH_2 scissoring bands near 1468 and 1423 cm^{-1} serve as additional confirmations of the increased tail ordering noted above.

Variation in a_0: ion-dipole interactions. There is one important mixed micelle case where it may be argued that the aggregate shape has been unequivocally established using FTIR spectroscopy, i.e. in mixtures of an amine oxide ($C_{14}AO$) with SDS (49). The $C_{14}AO$/SDS/water ternary phase diagram (for $\geq$ 70% water) contains just two pure phases: an isotropic liquid micellar phase (designation L_1) and a birefringent hexagonally packed liquid crystalline phase (designation H_1). The H_1 phase is composed of rodlike micelles packed into a hexagonal array (see Figure 1). A two phase area composed of the two pure phases (H_1+L_1) is also observed. The L_1 phase can be further divided into compositions yielding rodlike and spherical (globular) aggregates based on simple physical properties and rheological characteristics. The rodlike micelle phase is identified in the case of highly viscous samples, by the presence of streaming birefringence, and in lower viscosity samples, by the recoil of trapped air bubbles when swirling of the sample is discontinued. Viscosity is built into surfactant solutions by the formation of rodlike micelles, which can form entanglement couplings.

Methylene Tail Modes. Figure 12a shows the frequency of the composite symmetric CH_2 stretching band, and the zero shear viscosities (η_0) of mixed micelles formed between SDS and $C_{14}AO$ at 10% total surfactant (T = 20 °C). These compositions are all drawn from the L_1 phase. However, as the plot of the zero shear viscosities (η_0) shows, these micellar solutions are quite diverse, with a

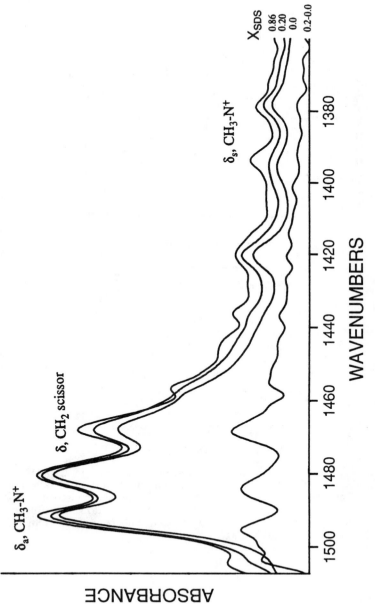

Figure 11: FTIR spectra of the CH_3-N^+ and CH_2 deformation region in mixed micelles of DTAC/SDS-d_{25}. From top, X_{SDS} = 0.86, 0.20, 0.0. The bottom spectrum is a difference spectrum with pure DTAC micelles subtracted from the X_{SDS} = 0.20 sample (T = 23°C). Reprinted from ref. 47. Copyright 1990 American Chemical Society.

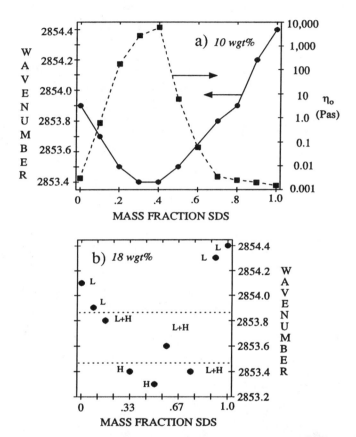

Figure 12: Plot of the symmetric CH$_2$ stretching frequency in C$_{14}$AO/SDS mixed micelles at 20°C. a) the total surfactant concentration is 10 wt.% and all samples are from the L$_1$ phase. Also plotted are the zero shear viscosities, η_0. b) the total surfactant concentration is 18 wt.% and the surfactant structure varies dramatically with composition (Reproduced with permission from ref. 49. Copyright 1990 Steinkopff Verlag).

variation in η_0 of seven orders of magnitude. The composite CH_2 frequency shows a composition-dependent minimum which corresponds closely to the large increase in solution viscosity. The composite v_s CH_2 frequencies at 18% total surfactant were also determined (Figure 12b). At this concentration, the $C_{14}AO/SDS$ mixtures pass through a number of different structures as a function of composition (i.e. micellar, 2-phase, hexagonal liquid crystal phases). The lowest frequencies are observed for the highly ordered H_1 phase, followed by the L_1+H_1 mixtures, and finally the L_1 solutions. As discussed above, the H_1 phase is composed of hexagonally packed long rodlike aggregates. The frequency of the composite v_s CH_2 band of this phase occurs at 2853.3-2853.4 cm^{-1}. By comparing Figures 12a and 12b, it is apparent that at a mass fraction of SDS of 0.3-0.4 (10% total surfactant, maximum in solution viscosity), the v_s CH_2 band occurs at 2853.4 cm^{-1}, which is equivalent to hexagonal phase samples. This similarity implies that the packing of the methylene tails is similar in the two instances, i.e. that the micelles are indeed rodlike, and that the rodlike micelles are simply precursors for the hexagonal phase, whereby at low volume fractions rodlike micelles are favored, while at higher volume fractions the hexagonal phase is favored. This is illustrated in more detail in Figure 1.

Headgroup Modes. An examination of the spectra of the mixed micelles in the fingerprint region can provide information about the interactions between the $C_{14}AO$ and SDS headgroups. Figure 16 shows a reference spectrum of micellar SDS (at 10% concentration) and a series of difference spectra, in the frequency range which includes the relatively intense v_{as} S-O and v_s S-O bands of the SDS headgroups.

The bottom seven difference spectra in Figure 13 show the change in shape of the broad v_{as} S-O band as the relative amount of $C_{14}AO$ in the mixed micelles (at 10% total surfactant concentration) is increased (8% SDS to 1% SDS). The appearance of the asymmetric bands in the difference spectra, in which the complete nulling of the symmetric S-O band near 1061 cm^{-1} was the subtraction criterion, indicates a significant increase in the relative intensity of the asymmetric band, due to a perturbation of a fraction of SDS headgroups by interactions with nearest neighbor $C_{14}AO$ molecules in the mixed micelles. The production of two distinct asymmetric S-O band shoulders in the difference spectra is again attributed to a "local site" symmetry lowering (an increase in ordering) of the SO_3 group by strong ion-dipole interactions with $C_{14}AO$, in a manner analogous to that observed in SDS/NaCl and SDS/DTAC mixtures.
The shape of the asymmetric S-O bands in the $C_{14}AO/SDS$ difference spectra change with composition. The higher frequency shoulder increases in relative intensity as the SDS content decreases. This change indicates that the dilution of SDS in $C_{14}AO$ results in a replacement of the sulfonate-sodium ion interactions with sulfate-amine oxide interactions, since the exact frequency and shape of the complex S-O band depends on the size and location of the hydrated counterion of the sulfate group (4). A weak band due to amine oxide is known to occur near 1200 cm^{-1}. The influence of this band on the v_{as} S-O bandshape can only be observed at the highest amine oxide concentration (mass fraction SDS = 1%) where a low frequency shoulder near 1200 cm^{-1} is observed in the difference spectra.
The shift of the symmetric S-O band from 1061.8 cm^{-1} for micellar SDS to 1060.1 cm^{-1} in the case of the mixed micelle containing the highest $C_{14}AO$ level is also consistent with a loss of interaction of the S-O group with sodium counterions (4,47).

Variation of v. The last mixed micelle case studied were mixtures in which the volume of the hydrophobic tails was varied by mixing monoalkyl and dialkyl cationic surfactants (DTAB/DDAB). The aggregate structures found as a function of composition are detailed in Table IV. They range from lamellar packed liquid

Table IV. Monoalkyl/dialkyl cationic surfactant mixtures: sample composition

Mole Fraction DTAB	P_{eff}^{1}	Phase
1.0	0.33	Micellar (L_1,spheres)
0.95	0.35	Micellar
0.90	0.38	Micellar
0.85	0.40	Micellar
0.80	0.43	Micellar
0.70	0.48	Micellar
0.65	0.50	Micellar (L_1, rods)
0.50	0.58	2-Phase ($L_\alpha + L_1$)
0.30	0.67	Lamellar Liquid Crystal
0.0	0.82	Lamellar Liquid Crystal (L_α)

[1] P_{eff} values calculated from $P_{eff} = X_{DTAB}P_{DTAB} + X_{DDAB}P_{DDAB}$. The values of the pure component packing parameters obtained by Evans et al.(15) from x-ray measurements.

crystalline phases (see Figure 1) to small spherical micelles. The association structure which forms in the DTAB/DDAB mixtures can be predicted using the packing parameter of Israelachvili *et al.* (*13,20*), wherein the mixture packing parameter is simply a mole fraction weighted sum of the two pure packing parameters. Thus, mixed micelles are expected at compositions ranging from X_{DTAB} = 0.65-1.0, with the rodlike micelles expected at the higher mole fractions of DDAB, i.e. $X_{DTAB} \approx 0.65$.

The surfactant-surfactant interaction parameter for these mixtures is zero, i.e. ideal mixing is observed (*50*). Ideal mixing implies that any composition dependent differences in packing which lead to the large changes in aggregate structure, are not sufficient to alter the free energy of micellization beyond what is predicted by ideal solution theory arguments (i.e. that the headgroup and tail packing are not radically different in the various structures).

Methylene Tail Modes. Upon examining the spectra, it is apparent that the frequencies of the CH_2 bands are higher in the spectrum of the spherical DTAB micelles (2926.2 and 2855.4 cm^{-1}) compared to those of the DDAB bilayers (2924.4 and 2854.3 cm^{-1}). This difference is expected, since the disorder (number of gauche conformers) of the methylene chains in the spherical micelles is expected to be greater than in the case of bilayers. The CH_2 stretching frequencies, for a given chain length, will rise as the number of gauche conformers is increased, and thus the disordering or fluidizing of the methylene tails is consistent with that found for phospholipid bilayers upon addition of a monoalkyl surfactant (*51*).

The ideality in surfactant-surfactant interactions found in the cmc plots is reflected in the composite ν_s CH_2 band frequencies plotted in Figure 14. The solid

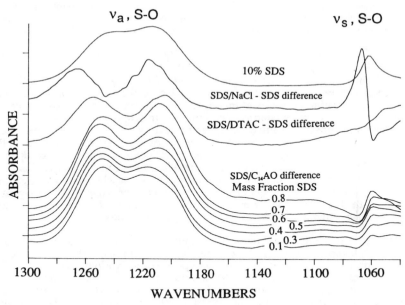

Figure 13: Difference spectra in the S-O stretching region for 10 wt.% $C_{14}AO/SDS$ mixtures (T = 20°C). Also pictured are difference spectra from DTAC/SDS mixed micelles, and SDS/NaCl mixtures, and the spectrum of SDS micelles in water (Reproduced with permission from ref. 49. Copyright 1990 Steinkopff Verlag).

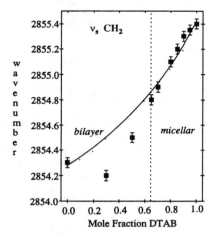

Figure 14: Frequency shifts of the composite symmetric CH_2 stretching band in actual and synthetic spectra of 0.3 M DTAB/DDAB mixtures (T = 25°C). The squares are the data points from actual spectra, while the line is from synthetic spectra.

line is derived from the frequency shift of the band in synthetic spectra generated using linear combinations of the two pure component spectra. The good agreement between the line and points for the micellar region agrees well with the concept of ideal mixing for the methylene chains. The slightly poorer agreement at higher mole fractions ($X_{DTAB} \leq 0.5$) indicates that the synthetic spectra are not able to predict the phase transition to bilayered structures adequately.

The shift of the composite v_s CH_2 band of the mixtures as a function of composition can be accounted for by the overlap of the CH_2 bands of the pure components, indicating nearly ideal mixing of the surfactants in the mixed micelles. This type of result strongly contrasts with the spectroscopic behavior of composite CH_2 bands in the spectra of mixed micelles formed by two surfactants which have attractive interactions in the headgroup region, for example DTAC and SDS. In the case of strongly interacting surfactants, the reduction in average headgroup area caused by electrostatic attraction results in crowding of the methylene chains. This increased crowding leads to a large reduction in the gauche conformer content (decreased CH_2 stretching frequencies) of the methylene chains. The frequency decrease for the highly nonideally mixing DTAC/SDS mixtures cannot be accounted for by spectral overlap, i.e. by synthetic spectra of these blends.

Headgroup Modes. Additional information about the changes in molecular packing with compositional changes can be obtained from inspection of the spectra in the fingerprint region. Figure 15 shows the spectra of the various mixtures.

There are two asymmetric CH_3-N^+ deformations, which are apparently nearly degenerate in the case of both solid CTAB (52) and dioctadecyldimethylammonium chloride (DODAC) in the gel or liquid crystalline phases of low water (21%) content (53), since only a single band between 1488 and 1485 cm^{-1} is observed. In contrast, the appearance of a band pair near 1490 and 1480 cm^{-1} is observed for a micellar CTAB solution, and also for mixtures of DTAB/DDAB. The appearance of two bands is thus correlated with a fully hydrated, disordered headgroup region. Rotation of the bulky cationic headgroups results in interactions between the methyl groups on neighboring molecules within the micelle, leading to inequivalency in packing of the methyl groups and hence an increase in frequency separation of the normally degenerate asymmetric deformation bands. It is also interesting to note that the frequency separation is lower in the spectra of DDAB bilayers (1488.9, 1482.0 cm^{-1}, $\approx$ 6 cm^{-1}) compared to that of DTAB micelles (1491.3, 1480.5, $\approx$ 11 cm^{-1}). The closer headgroup packing and decreased hydration in the case of the bilayers results in a more "solid-like" spectrum. Thus, changes in this band pair with mixed micelle composition are to be expected because of the necessary changes in headgroup packing which accompany such micelle shape changes. For the mixed micelle solutions the frequency separation shrinks from 10.8 cm^{-1} for pure DTAB to 9.7 cm^{-1} for $X_{DTAB} = 0.65$. As discussed above, the decreased frequency separation is correlated with a higher degree of ordering (possibly lower hydration and/or interaction with counterions) of the headgroups. This observation is consistent with the sphere to rod transition expected from changes in the packing parameter caused by compositional changes. As the DDAB content of the mixed micelles is increased, average headgroup spacing decreases, and the δ_{as} CH_3-N^+ bands are affected. Above $X_{DTAB} = 0.85$, there is good agreement between the actual and synthetic spectra, which implies that the mixed micelles are probably still nearly spherical at this composition.

In the case of the mixed bilayer sample ($X_{DTAB} = 0.3$), the frequencies of the bands are more similar to that of the bilayer than predicted by the synthetic spectra, suggesting, as the composite v_s CH_2 bands did, that at least 30 mole% DTAB can be incorporated into DDAB bilayers without severely destabilizing them. Conversely, addition of more than 15 mole% DDAB into DTAB micelles causes significant spectroscopic changes in the headgroup region, and hence suggests a micellar shape change.

Additional evidence for changes in headgroup packing at this mole fraction comes from the δ_s CH$_3$-N$^+$ (1398 cm^{-1}) band, and α-CH$_2$ scissoring (1420 cm^{-1}) bands (Figure 16). The δ_s CH$_3$-N$^+$ and δ α-CH$_2$ bands are found to be split into doublets in the spectra of mixtures containing greater than 15 mole% DDAB (X_{DTAB} = 0.85). The appearance of these doublets is not due to spectral overlap, as a comparison with the synthetic spectra in Figure 16 shows. The increased splitting of these headgroup bands suggests that the mixing of the surfactants in the headgroup region is very "nonideal" in contrast to the behavior of the methylene chains. The fact that the frequencies of the δ_s CH$_3$-N$^+$ and δ α-CH$_2$ bands are also not accounted for by the synthetic spectra for $X_{DTAB} \leq 0.85$ implies that the changes in the headgroup region packing are excellent indicators of the micelle shape change in these mixtures, even though the behavior of methylene tails is not.

The last major band in the fingerprint region is found near 1379 cm^{-1}, and can be assigned to the δ_s R-CH$_3$ (umbrella) mode. The frequency of this band in the spectra of the mixtures falls between that of DTAB (1379.6 cm^{-1}) and DDAB (1379.3 cm^{-1}). Figure 16 illustrates that this band does not differ between the real and synthetic spectra, i.e. its shape and frequency can be accounted for simply by spectral overlap. This lack of change with composition is not suprising since this group is the one farthest away from the headgroup region.

The spectroscopic data indicates that the increase in the volume of the tails, which is accomplished by adjustment of the composition of the mixed micelle, affects the packing of the headgroups and can cause a shape change. The mixing of the methylene tails is nearly ideal, however, as indicated by the smooth changes in the frequency of the CH$_2$ bands, which is due to a lack of any composition-dependent enhancement of hydrophobic interactions in this system. The strong electrostatic repulsion between the like-charged headgroups limits close approaches, and thus the overall free energy of micellization is not altered appreciably (W/RT = 0) with composition.

Temperature Dependence of Sphere to Rod Transitions.

SDS/NaCl Mixtures. The effect of temperature on the micelles formed in 70 mM SDS + NaCl solutions is presented below. Mazer *et al.* (*14*) have found that the aggregation number, N, is at a maximum for supercooled solutions below the critical micellization temperature (cmt), and decreases towards the value expected for a spherical micelle as the temperature is increased. The variations in N with temperature are dependent on the concentration of added electrolyte, with the rodlike micelles formed in high salt (0.6 M) showing large variations, and the spherical micelles formed in little (0.3 M) or no salt showing only small variations.

Figure 17 presents the results of an FTIR spectroscopic study of the effect of salt concentration on the cmt of 70 mM SDS. As Mantsch *et al.* (*4*) have shown in similar work with alkali hexadecylsulfates, the cmt can be related to the sudden increase in frequency of the v_s CH$_2$ bands as a function of temperature. The large increase in the *gauche* conformer content of the methylene chains of the surfactant tails as they "melt" at the cmt is responsible for this frequency shift. The effect of added salt is to raise the cmt of SDS, which is the cause of the "*salting out*" of this ionic surfactant at any given temperature. The cmt values, taken as the midpoint of the discontinuities in the frequency-temperature plots, agree well with those obtained by other means (*14,54*).

The absolute differences in the CH$_2$ stretching frequencies in the spectra of the coagel phases of the three different systems reflect the differences in packing of the methylene chains adopted at low temperatures. The CH$_2$ scissoring band is doubled in the spectrum of the SDS/0.3M NaCl system below the cmt, suggesting that an orthorhombic subcell is readily adopted. Examination of the S-O stretching bands of

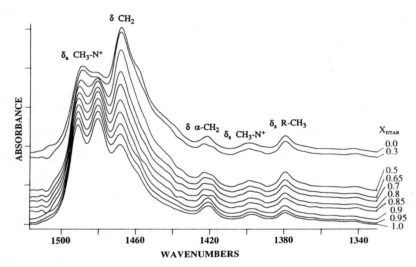

Figure 15: Fingerprint region spectra of 0.3 M DTAB, DDAB, and representative mixtures (T = 25°C).

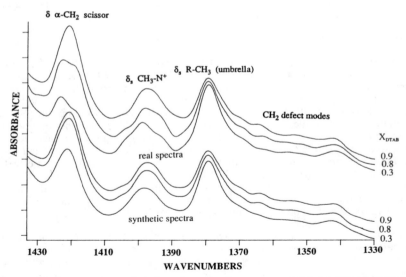

Figure 16: Comparison of real and synthetic spectra of 0.3 M DDAB/DTAB mixtures in the fingerprint region illustrating the band splitting in real spectra is not accounted for by spectral overlap (T = 25°C).

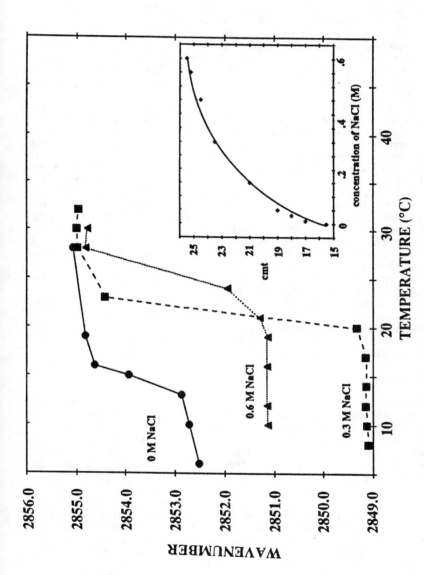

Figure 17: Temperature dependence of the symmetric CH$_2$ stretching mode at various electrolyte levels in 70 mM SDS+NaCl mixtures. Plotted in the inset are the critical micellization temperatures obtained as a function of added NaCl (*14,54*).

the coagels also reveals significant differences as a function of salt concentration. The S-O bands of anhydrous alkali hexadecylsulfates were significantly affected by the counterion type (4). Abrahamsson et al. (55) have suggested that crystalline SDS may exhibit a rather complicated unit cell involving four SDS molecules.

FTIR spectroscopy can provide useful information in the micelle phase, above the cmt, as illustrated in Figure 18. Note that there is a steady increase in frequency for the v_s CH_2 mode (Figure 18a) with increasing temperature, due to the increasing thermal disordering of the methylene chains. Also note that the frequency for the 0.6 M NaCl sample is always lower (at all temperatures) than the 0.3 M, which is lower than the sample without salt. These differences probably result from the increased ordering induced by the electrolyte in the headgroup region which promotes rod formation and methylene chain ordering.

Figure 18b and 18c illustrates changes in the symmetric S-O stretching mode and the CH_2 scissoring mode above the cmt. Overall, there is a small decrease in the symmetric S-O stretching band ($\approx$ 0.5 cm^{-1}), which is related to a small loss in bound counterions. Note that each salt level is affected similarly. Difference spectra in the S-O stretching region (Figure 19) as a function of temperature show that most of the difference occurs in the asymmetric mode, similar to that found for DTAC/SDS mixed micelles. This implies that the changes in micelle packing as a function of temperature occur between the surfactant molecules. One possible explanation is that the increased disorder for the methylene chains with increasing temperature leads to an increase in the effective headgroup area which affects the packing between surfactant molecules, thereby affecting the asymmetric mode to a greater extent. There are also slight decreases in counterion binding, which affect the symmetric mode, but not to the same extent. This is in contrast to the difference spectrum obtained as a function of electrolyte where the largest difference is seen for the symmetric mode. This is because the addition of electrolyte affects the location, angles, and numbers of Na$^+$ ions near the sulfate headgroups.

The CH_2 scissoring mode (Figure 18c) is extremely sensitive to the subcell packing, and suggests that just above the cmt, the methylene chains in the 0.6 M sample still pack in what is essentially a gel phase (ie. all-trans configuration). This packing becomes more liquid hydrocarbon-like with increasing temperature and decreasing electrolyte levels reflecting reduced interchain interactions.

Concluding Remarks

Rodlike micelles formed by a variety of methods have been studied by FTIR spectroscopy. A small decrease in the CH_2 stretching frequencies of the methylene tails is noted upon rod formation, and is linked to a decrease in the *gauche* conformer content. Difference spectra show that the decrease in CH_2 stretching frequencies are due to an increase in population of highly ordered methylene chains (similar in ordering to the *rotator* phase). This second type of methylene chain has been linked to molecules incorporated in the cylindrical portion of a spherocylindrical micelle. The increased ordering in the cylindrical portion of the spherocylinder is also noted by changes in the headgroup modes. The methylene tail ordering in rodlike micelles occurs preferentially near the headgroup, as noted by splittings in the headgroup, and α-CH_2 scissoring bands, and decreases towards the ends of the chain. The method by which rod micelle formation occurs (i.e. by addition of electrolyte, mixed micelles, or temperature) can be deduced by the use of difference spectra of headgroup bands. Overall, FTIR spectroscopy has shown itself to be an excellent technique for the study of micelle sphere to rod transitions.

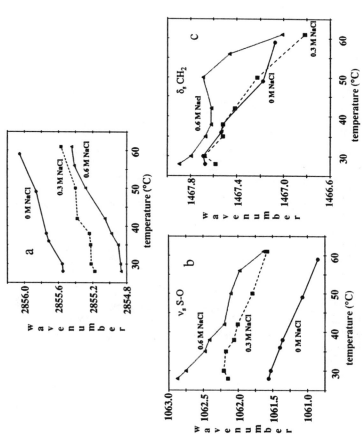

Figure 18: Plot of various modes as a function of temperature in 70 mM SDS + NaCl. a) symmetric CH_2 stretching mode b) symmetric S-O stretching mode c) CH_2 scissoring mode.

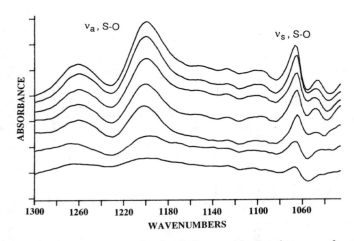

Figure 19: Difference spectra in the S-O stretching region as a function of temperature in 70 mM SDS/0.6 M NaCl solutions. The spectrum at 61°C has been subtracted in each case. Temperatures from top to bottom are 28, 30, 35, 38, 42, 50, 58 °C.

Acknowledgments

We acknowledge Daniel Webster for his skillful experimental techniques, and Dr. Henry Mantsch for stimulating discussions.

Literature Cited

1. Umemura, J.; Cameron, D.G.; Mantsch, H.H. *J. Phys. Chem.* **1980**, *84*, 2272.
2. Umemura, J.; Mantsch, H.H.; Cameron, D.G. *J. Colloid Interface Sci.* **1981**, *83*, 558.
3. Cameron, D.G.; Umemura, J.; Wong, P.T.T.; Mantsch, H.H. *Colloids Surf.* **1982**, *4*, 131.
4. Mantsch, H.H.; Kartha, V.B.; Cameron, D.G. In *Surfactants in Solution*; B. Lindman, B.; Mittal, K., Eds.; Plenum Press: New York, N.Y., 1984; Vol. 7, p 673.
5. Kawai, T.; Umemura, J.; Takenaka, T.; Kodama, M.; Seki, S. *J. Colloid Interface Sci.* **1985**, *103*, 56.
6. Yang, P.W.; Mantsch, H.H. *J. Colloid Interface Sci.* **1986**, *113*, 218.
7. Wong, P.T.T.; Mantsch, H.H. *J. Colloid Interface Sci.* **1989**, *129*, 258.
8. Kalyanasundaram, K.; Thomas, J.K. *J. Phys. Chem.* **1976**, *80*, 1462.
9. Cameron, D.G.; Kauppinen, J.K.; Moffatt, D.J.; Mantsch, H.H. *Appl. Spectrosc.* **1982**, *36*, 245.
10. Snyder, R.G.; Schachtschneider, J.H. *Spectrochimica Acta* **1963**, *19*, 85.
11. Holler, F.; Callis, J.B. *J. Phys. Chem.* **1989**, *93*, 2053.
12. Maroncelli, M.; Qi, S.P.; Strauss, H.L.; Snyder, R.G. *J. Am. Chem. Soc.* **1982**, *104*, 6237.
13. Israelachvili, J.N.; Mitchell, D.J.; Ninham, B.W. *J. Chem. Soc. Faraday Trans. I* **1976**, *72*, 1525.
14. Mazer, N.A.; Benedek, G.B.; Carey, M.C. *J. Phys. Chem.* **1976**, *80*, 1075.
15. Hoffmann, H.; Lobl, H.; Rehage, H.; Wunderlich, I. *Tenside Deterg.* **1985**, *22*, 290.
16. Zana, R.; Picot, C.; Duplessix, R. *J. Colloid Interface Sci.* **1983**, *93*, 43.
17. Weers, J.G.; *J. Amer. Oil Chem. Soc.* **1990**, *67*, 340.
18. Chang, D.L.; Rosano, H.L. In *Structure/Performance Relationships in Surfactants;* Rosen, M., Ed.; American Chemical Society Symposium Series: Washington, DC, 1984; Vol. 253, p 129.
19. Malliaris, A.; Binana-Limbele, W.; Zana, R.J. *J. Colloid Interface Sci.* **1986**, *110*, 114.
20. Evans, D.F.; Mitchell, D.J.; Ninham, B.W. *J. Phys. Chem.* **1986**, *90*, 2817.
21. Porte, G.; Poggi, Y.; Appell, J.; Maret, G. *J. Phys. Chem.* **1984**, *88*, 5713.
22. Casal, H.L. *J. Am. Chem. Soc.* **1988**, *110*, 5203.
23. Menger, F.M.; Jerkunica, J.M.; Johnston, J.C. *J. Am. Chem. Soc.* **1978**, *100*, 4676.
24. Halle, B.; Carlstroem, G. *J. Phys. Chem.* **1981**, *85*, 2142.
25. Cabane, B.; Zemb, T. *Nature* **1985**, *314*, 385.
26. Gruen, D.W.R. *J. Colloid Interface Sci.* **1981**, *84*, 281.
27. Dill, K.A.; Flory, P.J. *Proc. Natl. Acad. Sci. U.S.* **1981**, *78*, 676.
28. Szleifer, I.; Ben-Shaul, A.; Gelbart, W.M. *J. Chem. Phys.* **1985**, *83*, 3597, 3612.
29. Woods, M.C.; Haile, J.M.; O'Connell, J.P. *J. Phys. Chem.* **1986**, *90*, 1875.
30. Flory, P.J. *Statistical Mechanics of Chain Molecules;* Wiley: New York, N.Y., 1969.
31. Missel, P.J.; Mazer, N.A.; Benedek, G.B.; Young, C.Y.; Carey, M.C. *J. Phys. Chem.* **1980**, *84*, 1044.
32. Evans, D.F.; Ninham, B.W. *J. Phys. Chem.* **1983**, *87*, 5025.

33. Mazer, N.A.; Olofsson, G. *J. Phys. Chem.* **1982**, *86*, 4584.
34. Debye, P.; Anacker, E.W. *J. Phys. Colloid Chem.* **1951**, *55*, 644.
35. Leibner, J.E.; Jacobus, J. *J. Phys. Chem.* **1977**, *81*, 130.
36. Krathovil, J.P. *Chem. Phys. Letters* **1979**, *60*, 238.
37. Mantsch, H.H.; Martin, A.; Cameron, D.G. *Biochemistry* **1981**, *20*, 3138.
38. MacPhail, R.A.; Snyder, R.G.; Strauss, H.L. *J. Am. Chem. Soc.* **1980**, *102*, 3976.
39. Cameron, D.G.; Gudgin, E.F.; Mantsch, H.H. *Biochemistry* **1981**, *20*, 4496.
40. Lindman, B.; Wennerström, H. In *Topics in Current Chemistry*; Boschke, F.L., Ed.; Springer-Verlag: West Berlin, 1980; Vol. 87, p 1.
41. Dluhy, R.A.; Cameron, D.G.; Mantsch, H.H.; Mendelsohn, R. *Biochemistry* **1983**, *22*, 6318.
42. Donohue, J. *Soaps Cosmetics Chem. Spec.* **1985**, *1*, 32.
43. Gounoude, K.; *Soaps Cosmetics Chem. Spec.* **1985**, *12*, 52.
44. Scamehorn, J. *Phenomena in Mixed Surfactant Systems;* American Chemical Society Symposium Series: Washington, DC, 1986; Vol. 311, p 1.
45. Rubingh, D.N. In *Solution Chemistry of Surfactants*; Mittal, K.L., Ed.; Plenum Press: New York, N.Y., 1979; Vol. 1, p 337.
46. Rathman, J.F.; Scamehorn, J.F. *Langmuir* **1986**, *2*, 354.
47. Scheuing, D.R.; Weers, J.G. *Langmuir* **1990**, *6*, 665.
48. Filipović-Vinceković, N.; Škrtić, D. *Colloid Polym. Sci.* **1988**, *266*, 954.
49. Weers, J.G.; Rathman, J.F.; Scheuing, D.R. *Colloid Polym. Sci.* **1990**, in press.
50. Weers, J.G.; Scheuing, D.R. *J. Colloid Interface Sci.* **1990**, submitted.
51. Sinensky, M.; Kleiner, J. *Cellular Physiology* **1981**, *108*, 309.
52. Scheuing, D.R.; Reboa, P.F.; Weers, J.G. *Colloid and Surfaces*, **1990**, in press.
53. Umemura, J.; Kawai, T.; Takenaka, T.; Kodama, M.; Ogawa, Y.; Seki, S. *Mol. Cryst. Liq. Cryst.* **1984**, *112*, 293.
54. Nakayama, H.; Shinoda, K. *Bull. Chem. Soc. Jpn.* **1967**, *40*, 1797.
55. Abrahamsson, S.; Dahlén, B.; Löfgren, H.; Pascher, I. *Prog. Chem. Fats & Other Lipids* **1978**, *16*, 125.

RECEIVED August 9, 1990

Chapter 7

Alkyldimethylamine Oxide Surfactants

pH Effects on Aqueous Solutions

James F. Rathman and David R. Scheuing

The Clorox Company, 7200 Johnson Drive, Pleasanton, CA 94588

Fourier transform infrared spectroscopy was used to study pH effects on solutions of alkyldimethylamine oxide surfactants, C_nAO (molecular formula: $CH_3(CH_2)_{n-1}(CH_3)_2N{\rightarrow}O$ n=6, 8, 12). Results are presented for the neutral surfactant (high pH), the protonated (cationic) surfactant (low pH) and a 1:1 mixture (pH$\approx$pK$_a$). The frequency of the ν_s CH$_2$ band is used to monitor the monomer-to-micelle transition, as well as the effects of adjusting the solution pH at a given concentration. Band assignments are presented for the pH-sensitive ν_a C-N-O modes. For monomer solutions of C_8AO, the synthetic IR spectrum of the 1:1 mixture, constructed from a simple linear combination of the experimental spectra at high and low pH, agrees very well with the experimental spectrum, proving that monomer-monomer interactions are negligible in this particular system. Spectroscopic evidence for specific headgroup interactions between cationic and nonionic moieties in the 1:1 mixed micelle is observed.

Alkyldimethylamine oxides exist in aqueous solution in either nonionic or cationic (protonated) form. Depending on the length of the hydrophobic tail group, both forms may be surface active, and so the surfactant-related properties of these solutions can vary dramatically with pH. The effects of pH and electrolyte on the aggregation behavior of amine oxide surfactants have been studied by a variety of techniques, including light scattering (1-4), ^{13}C NMR (5), UV absorption (6), calorimetry (7,8), and rheology (9,10). These studies have shown that for solution pH near the pK$_a$, large rodlike micelles form, even at relatively low concentration, resulting in aggregation numbers and viscosities much greater than at either pH extreme.

An amine oxide surfactant solution can be modeled as a binary mixture of cationic and nonionic surfactants, the composition of which is varied by adjusting the pH. The cationic and nonionic moieties form thermodynamically nonideal mixed micelles, and a model has been developed which quantitatively describes the variation of monomer and micelle compositions and concentrations with pH and

0097–6156/91/0447–0123$06.00/0

total concentration (*11*). However, such a model does not provide a complete description of the system, since it does not address factors such as the structure of the micellar aggregate, the influence of changes in surfactant structure or concentration on the packing of members of micellar aggregates, or the effect of changing solution pH. Spectroscopic analyses can be used to this end, and IR techniques are especially useful for studying surfactant mixtures, since the different species can be monitored simultaneously. Surfactants such as amine oxides which exhibit acid-base equilibria offer the spectroscopist a unique opportunity for investigating binary surfactant mixtures. The ionic and nonionic species are structurally nearly identical in such mixed micelles, and hence the effect of electrostatic changes on micelle structure is more directly measured.

Experimental

Materials. Trimethylamine N-oxide dihydrate (98%) from Aldrich and N-dimethyldodecylamine N-oxide (97%) from Fluka Chemie were used as received. N-dimethylhexylamine N-oxide and N-dimethyloctylamine N-oxide were prepared by reaction of the corresponding tertiary amine with hydrogen peroxide (*12*). Both samples were isolated as crystalline solids and were >99% pure, based on acid/base titrations and spectrometric methods. Both samples were very hygroscopic. Reagent grade NaBr, 0.1 N and 2.0 N HCl were from J.T. Baker Chemical Co. and the all solutions were prepared using distilled and deionized water. The pH was monitored using an Orion Ross combination pH electrode and an Orion EA 940 meter.

Methods. All experiments were performed at 25°C. Critical micelle concentrations were determined using the maximum bubble pressure method on a SensaDyne 6000 surface tensiometer. Dry nitrogen was used as the gas source for the process and was bubbled through the solution at a rate of ~1 bubble/sec. Cmc's measured using the Wilhemy plate method were in agreement with those obtained from the bubble tensiometer; however, the bubble pressure method was used since it is less susceptible to error due to impurities and the nitrogen environment makes pH control easier.

IR spectra were obtained with a Digilab FTS 15/90 spectrometer equipped with a liquid nitrogen cooled wide band mercury-cadmium-telluride detector. Spectra of the aqueous solutions were obtained by the coaddition of 1024 scans at a nominal resolution of 4 cm^{-1}, using a triangular apodization function. All experiments were performed at 25 C, using a temperature controlled transmission cell (Harrick TLC-M25). Samples were placed between calcium fluoride windows, separated by a Mylar spacer at pathlengths of 15 or 25 μm. Sample temperatures were monitored with a thermocouple placed in contact with the edge of the windows. The spectra of the surfactants were obtained by subtracting the appropriate liquid and vapor water spectra, collected the same day at the same temperature as the sample solutions.

The symmetric CH$_2$ stretching band frequencies were found using a peak-finding program (*pkprogs*) developed by D.J. Moffatt at the National Research Council of Canada. This program utilizes a Fourier interpolation/zero slope algorithm which locates peak maxima. The spectra obtained in this study exhibited excellent signal to noise ratios, so the use of this or a center of gravity algorithm would be expected to yield similar results. The standard software supplied by Digilab for peak location was used for other bands. Cameron et al. (*13*) have demonstrated that frequency measurements with uncertainties of hundredths of a wavenumber or less can be routinely made, independent of instrument resolution. Precision experiments conducted in our laboratory on nine independent SDS solutions yielded a 95% confidence interval for the frequency of the symmetric CH_2 stretching band of $\pm$ .04 cm^{-1}. Similarly, measurement of successive spectra at various spectral resolutions, and on different days gave confidence intervals of $\pm$ .01-.02 cm^{-1}. These uncertainties reflect the errors in band frequency determination due to spectrometer drift, temperature control, and any unknown chemical heterogeneity of the samples. Before application of the peak finding programs, identical linear baseline corrections were applied to spectra undergoing direct comparisons; typically, baseline points were set at 3000 and 2800 cm^{-1} for CH_2 stretching bands. The spectral plots in this paper were produced on a Digilab 3240 computer. No noise reduction or smoothing algorithms have been applied to any of the spectra used in these studies.

Thermodynamic Considerations

Monomer Solutions. At surfactant concentrations less than the critical micelle concentration (cmc), all the surfactant is in monomeric form and the equilibrium between the protonated and neutral species of an alkyldimethyl amine oxide can be described by a classical dissociation constant, K_a:

$$CH_3(CH_2)_n - \overset{\overset{\displaystyle CH_3}{|}}{\underset{\underset{\displaystyle CH_3}{|}}{\overset{+}{N}}} - OH \quad \rightleftharpoons \quad CH_3(CH_2)_n - \overset{\overset{\displaystyle CH_3}{|}}{\underset{\underset{\displaystyle CH_3}{|}}{N}} \rightarrow O \quad + \quad H^+$$

$$K_a = a_{H+}\, a_o^{\,m} / a_+^{\,m} \approx a_{H+}\, C_o / C_+ \tag{1}$$

where a_{H+}, $a_o^{\,m}$, and $a_+^{\,m}$ are the activities of the hydrogen ion, and the neutral and cationic monomers, respectively, and C_o and C_+ represent molar concentrations of the monomers. As expressed in equation 1, it is generally assumed that the monomer activity and concentration are equal; i.e. surfactants in monomer form are assumed to obey Henry's Law, $(a_i^{\,m}/C_i) \rightarrow 1$ as $C_i \rightarrow 0$. For solution concentrations below the cmc, it is straightforward to show that concentrations of cationic and nonionic monomers are equal when the pH equals the pK_a (*11*).

Micellar Solutions. At concentrations above the cmc, where the amine oxide molecules are distributed between the micellar and monomeric forms, the situation is more complex since the degree of protonation of the micellar surfactant can be quite different than that of the monomeric surfactant. The mass balance for the surfactant is

$$Z\,(C_T) = Y\,(C_+ + C_o) + X\,(C_T - C_+ - C_o) \tag{2}$$

where C_T is the total (bulk) surfactant concentration, and X, Y, and Z are the micellar, monomeric, and overall degrees of protonation. As before, C_+ and C_o represent the monomer concentrations of the protonated and neutral species. C_T, Z, and pH are related by the solution charge balance

$$Z\,(C_T) + [H^+] = [OH^-] + [HCl] \tag{3}$$

A model has been developed which describes the variation of X, Y, C_+ and C_o with C_T and Z (*11*). The details of this approach relevant to this study may be summarized: 1) for $C_T \leq$ cmc, $Z=Y$; 2) for $C_T \gg$ cmc, $Z=X$; 3) for intermediate concentrations, Z, X, and Y are all different, except at $pH \approx pK_a$, where $Z=X=Y=0.5$. For this reason, results are presented here for $Z=0$, 0.5, and 1.0, since this permits the investigation of the entire concentration range without having to work with solutions in which the monomer and micellar compositions are not equal.

Results and Discussion

pK_a and Critical Micelle Concentrations. Values of the pK_a for the various amine oxides in aqueous solution are given in Table I. Also listed are critical micelle

Table I.   Values of the pK_a and Critical Micelle Concentrations for

Aqueous Solutions of Amine Oxides at 25 C

		critical micelle concentration (M)		
species*	pK_a	$Z = 0$	$Z = 0.5$	$Z = 1.0$
C_1AO	4.5	–	–	–
C_6AO	4.6	–	–	–
C_8AO	4.6	0.140	0.110	0.150
$C_{12}AO$	4.9	0.0020	0.0024	0.0060

* $C_nAO \equiv C_nH_{2n+1}(CH_3)_2N \rightarrow O$

concentrations for C_8AO and $C_{12}AO$ at $Z=0$ (nonionic form, $pH \gg pK_a$), $Z=1$ (cationic form, $pH \ll pK_a$) and $Z=0.5$ (1:1 mixture of nonionic and cationic forms, $pH \approx pK_a$). The cmc decreases with increasing alkyl chain length, and the cmc of the cationic form is greater than for the neutral surfactant. The cmc of the cationic form is considerably less than the cmc of conventional cationic surfactants having the same alkyl chain length; for example, the cmc of dodecyltrimethylammonium chloride in water is ~0.020 M, as compared to 0.0060 M for $C_{12}AO$ at $Z=1$. One explanation for this substantial difference   is that the hydroxy group of the protonated amine oxide hydrogen bonds with the aqueous phase surrounding the micelle, thereby reducing the repulsive forces between charged headgroups (14).

Spectroscopic results are presented for 0.08 M monomeric solutions of C_6AO and C_8AO, and 0.40 M micellar solutions of C_8AO and $C_{12}AO$. The discussion presented here is based on IR bands in two spectral regions: a) The "fingerprint" region (1500 to 950 cm^{-1}) contains the deformation modes of the methylene chain and the stretching modes of the surfactant headgroup. These band assignments are summarized in Tables II and III. b) The CH stretching modes, found at 3000 to 2800 cm^{-1}, are summarized in Table IV.   Details of these assignments are discussed below.

Trimethylamine Oxide Spectra. Assignment of the major IR bands associated with the amine oxide headgroup were made based on aqueous solution spectra of trimethylamine oxide (Figure 1), the spectrum of solid $C_{12}AO$ (Figure 2), and

Table II.   IR Band Assignments of the Deformation Modes of
Aqueous Solutions of Amine Oxides at 25 C

	Z	δ_a CH_3-N	δ_s CH_3-N	δ $\alpha-CH_2$	δ CH_2	δ_s $R-CH_3$
C_1AO	0.0	1466, 1482	1404	–	–	–
	0.5	1465, 1483	1407	–	–	–
	1.0	1462, 1486	1410	–	–	–
C_6AO	0.0	1460, 1476	1403	1431	1470	1383
(monomer)	0.5	1460, 1475	1404	1432w	1470	1383
	1.0	1460, 1479	1410, 1388	1433w	1469	1383
C_8AO	0.0	1460, 1475	1403	1430	1471	1381
(monomer)	0.5	1460, 1477	1404	1431w	1470	1381
	1.0	1459, 1479	1410, 1388	1433w	1469	1381
C_8AO	0.0	1459, 1475	1403	1430	1470	1380
(micelle)	0.5	1458, 1477	1404	1434w	1469	1380
	1.0	1458, 1479	1410, 1388	1437w	1469	1380
$C_{12}AO$	0.0	1459, 1475	1402	1430	1468	1379
(micelle)	0.5	1458, 1477	1404	1437w	1468	1379
	1.0	1457, 1480	1409, 1388	1436w	1468	1379

δ_s = symmetric deformation

δ_a = antisymmetric deformation

(w) = band intensity is relatively weak

literature data for alkyltrimethylammonioum halides and alkylpyridinium oxides. FT-IR spectra of 0.2 M trimethylamine oxide (C_1AO) at various pH's are shown in Figure 1. A band of moderate intensity at 1240 cm^{-1} in solutions at high pH shifts to 1254 cm^{-1} upon acidification; the pH-dependence suggests the assignment of this band to a vibrational mode involving primarily C-N-O stretching. The formation of C-N$^+$-OH at low pH increases the ionic character of the N-O bond. The resultant increase in the force constant of the bond is responsible for the band shift toward higher frequency upon protonation of the headgroups. The increased relative intensity of this band at low pH is also consistent with increased ionic character of the N-O linkage. This assignment is supported by previous work with pyridine N-oxides (15-17), in which a band near 1250 cm^{-1} was assigned to N→O stretching. The addition of electron-withdrawing substituents to the pyridine ring resulted in higher frequencies for this band, as did the addition of substituents (e.g. amide groups) capable of forming intra-molecular hydrogen bonds to the N→O group (i.e. N→O--H).

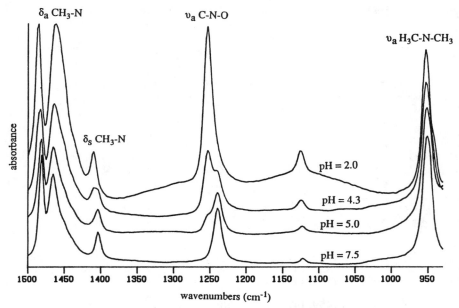

Figure 1. Effect of pH on the IR spectrum of 0.20 M trimethylamine oxide (C_1AO) in water at 25°C.

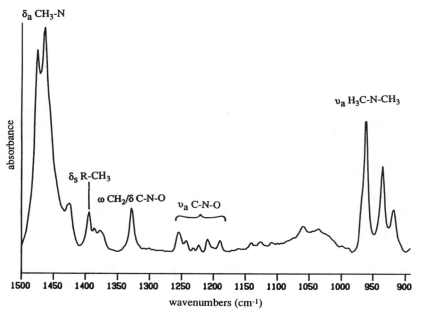

Figure 2. IR spectrum of solid dimethyldodecylamine oxide ($C_{12}AO$) at 25°C.

Table III.   IR Band Assignments of the Hydrophilic Group Stretching Modes for Aqueous Solutions of Amine Oxides at 25 °C

	Z	ν_a				$H_3C - N - CH_3$
		C– N – O				
C_1AO	0.0	–	1240	–	–	951
	0.5	–	1240	1254	–	952
	1.0	–	–	1254	–	953
C_6AO	0.0	1196	–	1225	1245	weak
(monomer)	0.5	1197	–	1227	1245	"
	1.0	1198	1205	1230	1245	"
C_8AO	0.0	1194	1206	1226	1250	"
(monomer)	0.5	1194	1208	1228	1248	"
	1.0	1194	1213	1230	1244	"
C_8AO	0.0	1194	1205	1227	1249	972
(micelle)	0.5	1192	1203	1229	1243	973
	1.0	1193	1212	1232	1244	974
$C_{12}AO$	0.0	1196	1223	1235	1249	969
(micelle)	0.5	1190	1220	1233	1242	969
	1.0	1199	1209	weak	1242	974

In an early IR study of alkylamine oxides (*18*), the band near 950 cm^{-1} was assigned to the N→O group; however, we assign this band to C-N-C stretching by analogy with similar bands observed in the spectra of alkyldimethyl- and trimethyl-ammonium compounds such as dodecyltrimethyl- and didodecyldimethyl ammonium bromide, phosphatidyl choline lipids, and tri- and tetramethyl amine salts (*19-22*). For C_1AO, the frequency of the 950 cm^{-1} band increases only slightly upon acidification due to the higher bond force constant for C-N$^+$ compared to C-N. Unlike the 1250 cm^{-1} band, there is no substantial change in the intensity of the 950 cm^{-1} band with changing pH, suggesting that this vibrational mode does not involve substantial motion of the charged groups.

The symmetric deformation of the CH_3 groups in CH_3-N is assigned to the weak band near 1404 cm^{-1}, which exhibits a shift to 1410 cm^{-1} upon protonation. There are two asymmetric deformation modes for the CH_3-N groups, at 1482 and 1466 cm^{-1}, which shift in opposite directions upon protonation. The appearance of two bands is a consequence of the phasing of the asymmetric deformations of the CH_3 groups attached to a common nitrogen atom. Iso-propyl and iso-butyl groups provide a useful analogy since FT-IR spectra of compounds containing these groups also exhibit two CH_3 deformation bands due to in-phase and out-of-phase motions of the groups attached to a common carbon atom.

Table IV. IR Band Assignments of the Hydrophobic Group Stretching Modes for Aqueous Solutions of Amine Oxides at 25° C

	Z	v_s CH_2	v_a CH_2	v_a CH_3
C_6AO (monomer)	0.0	2867.1	2935.5	2964.4
	0.5	2867.4	2935.9	2965.3
	1.0	2868.4	2937.8	2967.6
C_8AO (monomer)	0.0	2863.6	2934.5	2961.8
	0.5	2863.9	2934.8	2962.0
	1.0	2864.3	2935.8	2963.9
C_8AO (micelle)	0.0	2860.5	2931.5	2960.1
	0.5	2859.9	2931.6	2959.0
	1.0	2860.7	2931.6	2959.7
$C_{12}AO$ (micelle)	0.0	2855.1	2925.5	2956.9
	0.5	2854.8	2924.9	2956.6
	1.0	2855.1	2925.7	2957.2

v_s = symmetric stretching

v_a = antisymmetric stretching

Spectra of Monomeric Surfactant Solutions. The next step was to study alkyldimethylamine oxides having methylene chains of sufficient length such that the molecule is surface active. The FT-IR spectra of monomer solutions of C_8AO at various degrees of protonation (Z) are shown in Figure 3. These spectra are much more complex than those of C_1AO due to the introduction of deformation modes of the CH_2 and the terminal CH_3 groups.

The frequency of the v_s CH_2 bands in the spectra of monomer solutions of C_6AO and C_8AO are listed in Table IV. For a given value of Z, the v_s frequencies of monomeric C_8AO are significantly lower than those of C_6AO. The CH_2 groups near the surfactant headgroup, especially the α CH_2 group, experience a different environment than those farther away for two reasons: 1) inductive effects of the highly polar N-O group, which becomes ionic upon protonation; 2) the portion of the methylene chain near the headgroup is exposed to the hydration shell of the highly hydrated N-O groups. Both inductive and "solvent" effects would result in higher frequencies for CH_2 groups near the headgroup. As the length of the methylene chain is increased, the fraction of CH_2 groups influenced by the headgroup decreases, and hence the v_s and v_a CH_2 frequencies decrease, reflecting the increased contribution of unperturbed CH_2 groups to the overall band profile. The CH_2 deformation (scissoring) band is assigned to the region between 1472 and 1468 cm^{-1}. The symmetric deformation of the terminal methyl group on the alkyl chain gives rise to a new band near 1384 cm^{-1} (C_6AO) and 1381 cm^{-1} (C_8AO); as expected, this band is relatively insensitive to pH.

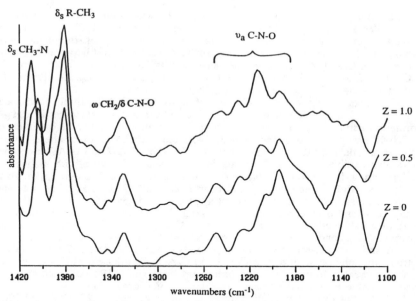

Figure 3a. The fingerprint region of monomeric solutions (0.08 M) of dimethyloctylamine oxide (C_8AO) at mole fractions of the protonated species (Z) equal 0, 0.5 and 1.0.

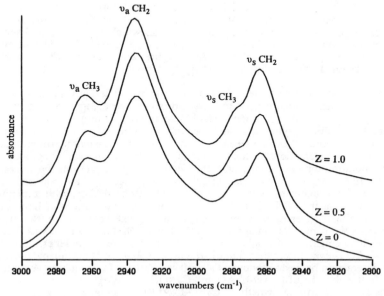

Figure 3b. The CH stretching region of monomeric solutions (0.08 M) of C_8AO at Z equal 0, 0.5 and 1.0.

Despite the complexity of these spectra, the headgroup bands of surfactant amine oxides are similar to those of C_1AO. The symmetric CH_3-N deformation is assigned to the band at 1403 cm^{-1}, which shifts to a higher frequency upon protonation. The asymmetric CH_3-N deformation band is found between 1474 and 1481 cm^{-1} for C_6AO and C_8AO, in the same vicinity (1466 to 1482 cm^{-1}) observed for C_1AO. This band is not as well defined for C_6AO and C_8AO, due to the significant overlap with the CH_2 scissoring bands.

The spectra of monomer solutions of C_6AO and C_8AO exhibit bands near 1464 and 1456 cm^{-1}. Based on the C_1AO results, a second asymmetric CH_3-N deformation band is expected in this region. Since an asymmetric deformation of the CH_3-R group is also expected near 1456 cm^{-1}, the overlap of these modes precludes any detailed discussion of these bands. The C_6AO and C_8AO monomer spectra also exhibit a peak of moderate intensity near 1435 cm^{-1}, with a shoulder peak appearing in spectra of solutions at high pH. The α-CH_2 group of alkyl carboxylates exhibit a scissoring band near 1420 cm^{-1} (*23*); alkyltrimethyl- and dialkyldimethyl- ammonium salts exhibit a similar band near 1430 cm^{-1} (*20,21*), as do fatty acids and triglycerides (*24,25*). Assignment of the 1435 cm^{-1} band in the spectra of alkyldimethylamine oxides to the α-CH_2 group is thus reasonable. The frequency of this band is sensitive to protonation of the headgroup, as expected.

A series of pH-sensitive bands appear between 1190 and 1270 cm^{-1} in the spectra of monomeric solutions of C_6AO and C_8AO. In the case of C_8AO (Figure 3a), as Z increases from 0 to 1, the prominant band near 1194 cm^{-1} decreases in intensity relative to a shoulder near 1206 cm^{-1}, which grows into a separate band near 1213 cm^{-1} at Z=1. The relative intensities of bands near 1227, 1248 and 1265 cm^{-1} also decrease with increasing Z. We assign this group of bands to vibrational modes of the C-N-O headgroups. The appearance of multiple C-N-O stretching bands may be rationalized by the presence of both H_3C-N-O and H_2C-N-O groups. In addition, CH_2 wagging motions are known to give rise to a series of weak bands in this frequency range in the spectra of disordered hydrocarbons (*26*). Coupling of these modes with the asymmetric C-N-O stretching modes may be expected. As the alkyl chain length is increased, the complexity and overlap of this multiplet increases; e.g., in the spectra of $C_{12}AO$ solutions, which are discussed later, two broad bands near 1198 and 1245 cm^{-1} are found. The well-defined multiplet is restored in the spectrum of solid $C_{12}AO$ (Figure 2), when the disorder of the methylene chains is largely eliminated by crystal formation.

Monomer-Micelle Interactions. As discussed above, it is convenient to treat an amine oxide solution as a binary surfactant mixture, the composition of which is varied by adjusting the solution pH. A common assumption made in the development of thermodynamic models describing monomer-micelle equilibrium is that there are no interactions between monomers. For a surfactant mixture, there are two types of interactions which must therefore be proved neglible: 1) nonionic-nonionic or ionic-ionic, and 2) ionic-nonionic. If present, the interaction between like-charged monomers might be expected at all values of Z. Cationic-nonionic interactions would be observed only at 0<Z<1, with a maximum effect expected near Z=0.5. For C_5 to C_7 alkanoates, Umemura *et al* have shown that premicellar aggregation is reflected in a gradual decrease in the υ_a-CH_2 frequency as the concentration approaches the cmc (*27*). No evidence was found of premicellar aggregation of n-alkanoates having 8 or more methylenes. Since the cmc of C_8AO is even lower than that of sodium octanoate, it is reasonable to assume that the interaction between monomers at Z=0 and Z=1 for C_8AO is also neglible.

In the absence of C_8AO-C_8AO and C_8AOH^+-C_8AOH^+ interactions, it is still important to determine whether C_8AO-C_8AOH^+ monomer interactions are significant. Such interactions would invalidate the assumption that the monomer pseudophase obeys Henry's Law (equation 1). The spectrum at Z=0.5 in Figure 3 should be a simple linear combination of the spectra at Z=0 and Z=1 if the nonionic and cationic monomers are truly noninteracting. Figure 4 compares experimental and synthetic spectra at Z=0.5. The experimental spectra at Z=0 and Z=1 were normalized to the intensity of the δ_s CH$_3$-R band near 1381 cm^{-1}, and then added with equal weighting to produce the synthetic Z=0.5 spectrum. The frequencies of some of the major bands in the synthetic spectrum are compared with the experimentally observed values in Table V.

Table V.   Comparison of Selected Band Frequencies at Z=0.5 from Synthetic and Actual Spectra for a Monomer Solution (0.08 M C$_8$AO) and a Micellar Solution (0.40 M C$_{12}$AO)

mode	0.08 M C$_8$AO		0.40 M C$_{12}$AO	
	synthetic	actual	synthetic	actual
δ_s R-CH$_2$	1381	1381	1379	1379
δ CH$_2$	1470	1470	1468	1468
υ_s CH$_2$	2864	2864	2855	2855
ω CH$_2$	1330	1330	1330	−
υ_a C-N-O	1194	1194	1190	1197
	1208	1209	1220	1228
	1228	1229	1233	−
	1248	1248	1242	1243

The overall aggreement between the synthetic and experimental spectra is remarkably good, providing support for the stated assumption. Slight discrepancies between the synthetic and real spectra may be attributed to changes in absorptivity of some of the bands with pH, which were ignored in this simple analysis.

The Monomer-to-Micelle Transition. Surfactants having relatively high cmcs facilitate FT-IR investigations of the monomer to micelle transition, since monomer solutions can be prepared at concentrations that readily provide acceptable signal/noise ratios. However, if the cmc is too high, it is not always possible to prepare concentrated, isotropic micellar solutions for comparison with those of the monomer. Of the amine oxide surfactants studied here, C$_8$AO is the best choice for examining this transition since its cmc (0.11 M to 0.15 M) is in an appropriate range.

IR spectra of micellar solutions of C$_8$AO at Z=0, 0.5 and 1.0 are shown in Figure 5. As illustrated in Figure 6, the υ_s CH$_2$ band is especially useful for investigating the effects of methylene chain length and micellization. For a given value of Z, the frequency of the υ_s CH$_2$ band decreases with increasing chain length. For a given surfactant monomer, the frequency of this band is higher for Z=1

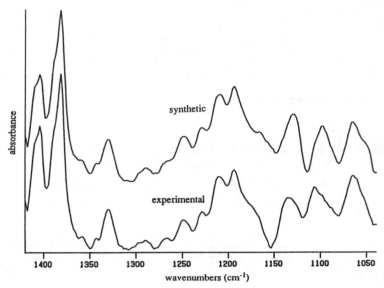

Figure 4. Comparison of synthetic and experimental spectra of a monomeric solution (0.08 M) of C_8AO at Z equal 0.5.

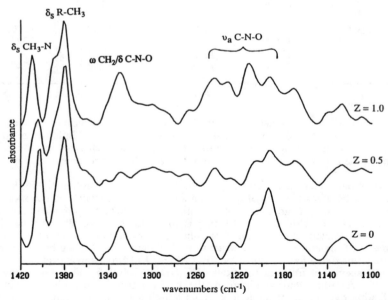

Figure 5. IR spectra of micellar solutions (0.40 M) of C_8AO at Z equal 0, 0.5 and 1.0.

compared to Z=0, the difference being greater for C_6AO than for C_8AO. Both of these trends are a consequence of the inductive effect that the polar headgroup (N→O) has upon the methylene chain, which becomes more pronounced at Z=1 when the monomer is protonated (producing N+–O–H). Increasing the methylene chain length reduces the fraction of CH_2 groups influenced by the headgroup, and therefore the band is observed at successively lower frequencies as the chain length is increased.

Of the frequency shifts reported in this work, the largest is the change in the v_s CH_2 band noted for the monomer to micelle transition of C_8AO. The frequency of the v_s CH_2 band in the spectrum of a surfactant molecule incorporated in a micelle is lower than that of the same surfactant as an unassociated monomer in solution (27), due primarily to the large decrease in the contact between the aqueous solvent and the hydrophobic methylene chain. This solvent effect was also observed in this study for C_8AO, as shown in Figure 6 and Table 4.

The changes in frequency for the headgroup modes are relatively small, in contrast to the large shifts in the CH stretching bands associated with the monomer to micelle transition. As shown in Tables 2 and 3, the CH_3–N deformation and the C–N–O stretching frequencies show only slight differences between the monomer and micellar solutions of C_8AO at Z=0. The same is true at Z=1, indicating that, for a solution containing a single amine oxide species, the environment surrounding the headgroups in the monomer and micellar states is apparently quite similar. At Z=0.5, several of the headgroup bands do show large differences between monomer and micelle, suggesting that the   formation of mixed micelles has a pronounced effect upon the headgroups.

Solutions at concentrations above the cmc may contain significant concentrations of both monomeric and micellar surfactant. Previous researchers have used FT-IR to investigate monomer-to-micelle transitions (27) and gel-to-liquid crystal transitions of lipid bilayers (28,29). These studies have demonstrated that, in the case of such two-state transitions, linear combinations of the infrared spectra of the initial and final states characterize the spectra of the intermediate states, where both forms coexist. However, changes in band frequency or width are not necessarily linear with the extent of the transition. Linear combinations of two highly overlapped Lorenztion bands can give rise to non-linear shifts in the band frequency and width (27-29).

As shown in Table 1, the cmc for C_8AO is 0.14 M, 0.11 M and 0.15 M for Z=0, 0.5, 1.0, respectively. A ssuming that the monomer concentration is constant above the cmc, a 0.40 M solution of C_8AO at Z=0 therefore has 65% of the surfactant in micellar form and 35% as monomer. The fraction of the surfactant in micellar form rises to 72% at Z=0.5, and falls back to 62% at Z=1.0. Subtraction of the monomer contribution from the spectra in Figure 5, yields spectra which, theoretically, are of the micellar surfactant only. The spectra in Figures 3 and 5 were first normalized to the intensity of the δ_s R–CH_3 band. Synthetic spectra were then obtained by subtracting the appropriately scaled monomer spectrum in Figure 3 (e.g. 0.35 for Z=0) from the corresponding spectrum in Figure 5. The resulting synthetic spectra are shown in Figure 7.   The key assumptions made in these calculations are that the monomer concentration at any given Z is constant at concentrations greater than the cmc, and that the monomer activity is unaffected by the presence of micelles (i.e. the monomer spectrum is the same for concentrations below and above the cmc).   Both of these assumptions are good first approximations, although, especially for solutions containing an ionic surfactant, the monomer concentration is expected to decrease somewhat with increasing concentration above the cmc.

The band positions of the headgroup modes for the synthetic spectra in Figure 7 are the same as observed for the 0.40 M C_8AO solution (Figure 5);

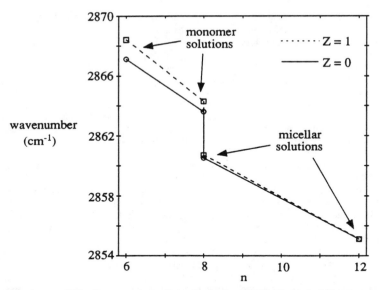

Figure 6. ν_s CH$_2$ frequencies of nonprotonated (Z=0) and fully protonated (Z=1) C$_n$AO amine oxides. Data are shown for monomer solutions (0.08 M) of C$_6$AO and C$_8$AO and micellar solutions (0.40 M) of C$_8$AO and C$_{12}$AO.

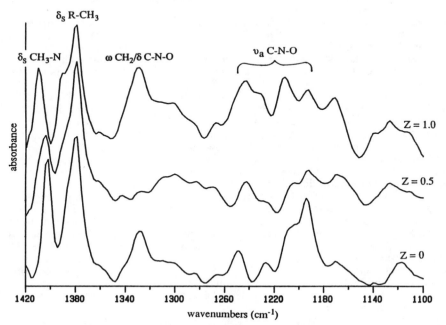

Figure 7. Synthetic spectra of micellar solutions (0.40 M) of C$_8$AO at Z equal 0, 0.5, 1.0.

however, differences are seen in the tailgroup bands, as illustrated in Figure 8. The v_s frequencies for the surfactant in micellar form (in the synthetic micelle spectra) are lower than those for the 0.40 M solution, which are slightly higher due to the presence of a significant concentration of surfactant monomer. The subtraction of the monomer spectrum from the spectrum of a solution above the cmc is a useful method for reducing spectroscopic contributions from the monomer, when present at significant levels in the system under study.

Another important feature of the data shown in Figure 8 is that v_s CH_2 for the micellar system is lower at Z=0.5 than at either Z=0 or Z=1; this trend is not observed for the monomer solutions (C_6AO or C_8AO), and will is discussed further below.

Micellar Solutions. IR spectra of micellar solutions of $C_{12}AO$ are shown in Figure 9. From Table 4, at a given value of Z the frequencies of v_s and v_a CH_2 for micellar $C_{12}AO$ are found to be lower than in the case of micellar C_8AO. This is the same trend discussed above for increasing the alkyl chain length of the monomer forms, and the same reasoning applies for the micelle: a combination of inductive and solvent effects result in higher stretching frequencies for the CH_2 groups near the headgroup. As the methylene chain length is increased, a smaller fraction of CH_2 groups are affected, and thus the band appears at a lower frequency.

Unlike monomer solutions, the frequency of the CH stretching modes for micellar solutions do not vary monotonically with Z, but rather go through a minimum. The trend noted in Figure 8 for the v_s CH_2 band is also seen for the v_a CH_3 band for both C_8AO and $C_{12}AO$, as shown in Figure 10. These results suggest that the hydrophobic methylene chains in the interior of the mixed micelle at Z=0.5 are packed in a way which reduces contact with the aqueous phase even more than at Z equal 0 or 1. The frequency minimum is also consistent with a decrease in the gauche/trans conformer ratio, i.e., a partial straightening of the methylene chain tails (*22*), at Z=0.5. Such behavior is consistent with the amine oxide micelles being roughly spherical when only nonionic or cationic moeities are present, and becoming more rodlike at Z=0.5

One of the most striking features of the C_8AO and $C_{12}AO$ micellar spectra is that the band near 1330 cm^{-1} at Z=0 and Z=1 is totally absent at Z=0.5. The weaker, broader band near 1300 cm^{-1} which is more readily detected at Z= 0.5 can be readily assigned to the wagging of CH_2 groups of so - called "kink" defects (a gauche-trans-gauche' conformation) common to disordered methylene chains (*30*). Comparing Figures 3, 5 and 9, the disappearance of the 1330 cm^{-1} band is observed only for micellar solutions. Wagging bands (ω CH_2) are known to occur in this region. Since the surfactant methylene chains in the micelle are fairly liquid-like, wagging band progressions characteristic of a solid or gel state are not expected; however, specific gauche defects of the methylene chain also exhibit bands in this region (*26*). We believe the band at 1330 cm^{-1} to be a combination of the α-CH_2 wagging mode with a deformation (bending) mode of the C-N-O group. The absence of this band in micellar systems at Z=0.5 suggests that the C-N-O bending is absent, or at least greatly restricted. This would be expected if the headgroups are much closer in the mixed micelle, which is consistent with the v_s CH_2 results indicating the packing of methylene chains in the mixed micelle is also tighter than for the single-component micelles. This interpretation is reasonable in light of results for more conventional types of surfactant mixtures, such as dodecyltrimethyl-ammonnium chloride/sodium dodecyl sulfate, where the ordering of the methylene chains and headgroups increased dramatically as the mixed micelle composition approached equimolar (*31*).

The appearance of rodlike micelles at Z=0.5 also influences the headgroup modes; as shown in Table 3, for example, the frequency of the v_a C–N–O bands is

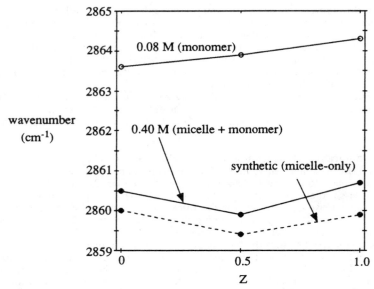

Figure 8. Comparison of the ν_s CH$_2$ frequencies of C$_8$AO from experimental spectra of 0.08 M and 0.40 M solutions, and synthetic micelle-only spectra.

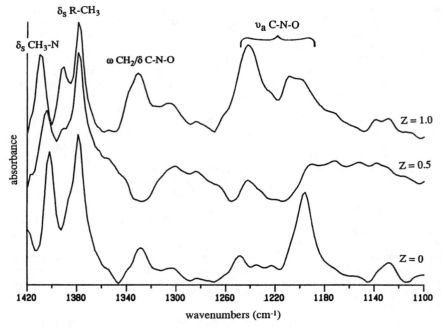

Figure 9. IR spectra of micellar solutions (0.40 M) of C$_{12}$AO at Z equal 0, 0.5 and 1.0.

generally lower at Z=0.5 than at either extreme. Again, this trend is only observed for micellar solutions. The unique headgroup environment of micelles at Z=0.5 is indicated by the shifts toward lower frequency and broadening of the C-N-O stretching bands. Weakening of the N-O bond strength through extensive hydrogen - bonding between "protonated" and neutral molecules at Z=0.5 may be responsible for the shifts observed. It is thus apparent from the spectra that both the packing (effective headgroup area) and the electrostatic interactions between headgroups at the micelle surface are quite distinct for the mixed micelles formed at Z=0.5

The discussion thus far has focused on the variation of band positions, since the complexity of the spectra, especially for the micellar systems, makes quantitative discussion of the peak intensites or bandwidths difficult. It is, however, possible to gain several useful insights by qualitatively comparing the relative peak intensities. The cationic/nonionic mixed micelle at Z=0.5 is expected to be thermodynamically nonideal, meaning that interactions in the mixed micelle differ from interactions in the single-component micelles. Thus, unlike the synthetic monomer spectrum for C_8AO at Z=0.5 (Figure 4), the addition of the Z=0 and Z=1 spectra would not be expected to yield a synthetic spectrum that is the same as the experimental result at Z=0.5 for micellar solutions. The actual and synthetic spectra for 0.40 M $C_{12}AO$ at Z=0.5 are compared in Figure 11. With regards to bandwidths and the relative intensities of adjacent peaks, the calculated and actual spectra are similar in the CH stretching region (3000-2800 cm^{-1}) and very different in the region containing the headgroup modes (1400-1100 cm^{-1}). These data again indicate that the formation of mixed micelles results in very different interactions between headgroups than exist in the single-component micelles. Also, the effect of mixing on the hydrophobic groups inside the micelle is relatively small in comparison to those in the headgroup region.

Summary

FT-IR provides a wealth of information for both monomer and micellar surfactant solutions. The v_s CH_2 band can be used to monitor changes in the environment of the methylene chain. The frequency of this band decreases when the chain length is increased, due to inductive effects of the polar C-N-O group on the vibrational modes of the CH_2 groups near the headgroup. This frequency is also lower for micellar surfactant compared to monomer, and for mixed micelles compared to single-component micelles; these trends are a result of ordering of the methylene chains in the micelle and reduced contact with the water phase. The v_a C-N-O multiplet exhibits an increase in the relative intensity of the bands at higher frequency as the surfactant is protonated. The C-N and N-O bond force constants are greater for the $C-N^+$-OH species than for the neutral C-N→O. For monomer solutions of C_8AO, the synthetic IR spectrum of the 1:1 mixture, constructed from a simple linear combination of the experimental spectra at high and low pH, is in excellent agreement with the experimental spectrum. This provides conclusive support for the common assumption that monomer-monomer interactions are negligible in this type of surfactant system. Conversely, spectroscopic evidence for specfic headgroup interactions between cationic and nonionic moieties in the 1:1 mixed micelle are observed. This indicates that the major source of the thermodynamic nonidealities associated with the formation of mixed cationic/nonionic micelles arises from headgroup interactions that are much different in the mixed micelle than in either single-component micelle.

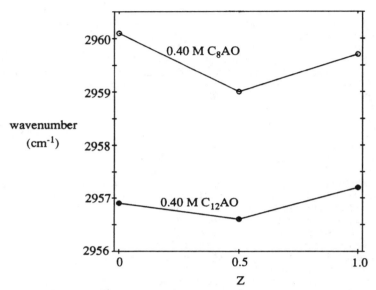

Figure 10. υ_a CH$_3$ frequencies of micellar solutions (0.40 M) of C$_8$AO and C$_{12}$AO at Z equal 0, 0.5, and 1.0.

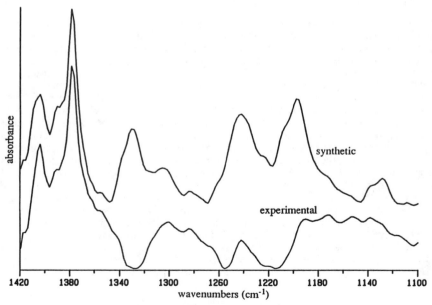

Figure 11a. Comparison of the fingerprint region of synthetic and experimental spectra of 0.40 M C$_{12}$AO at Z equal 0.5.

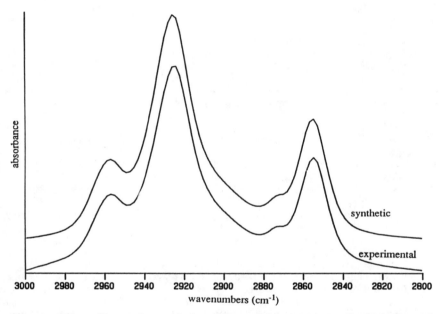

Figure 11b. Comparison of the CH stretching region of synthetic and experimental spectra of 0.40 M $C_{12}AO$ at Z equal 0.5.

Acknowledgments

The C_6AO and C_8AO surfactants were synthesized and purified by Lafayette Foland. The authors also wish to acknowledge their useful conversations with Jeff Weers.

Literature Cited

1. Hermann, K.W. *J. Phys. Chem.* **1962**, *66*, 295.
2. Ikeda, S.; Tsunoda, M.; Maeda, H. *J. Colloid Interface Sci.* **1979**, *3*, 448.
3. Imae, T.; Ikeda, S. *Colloid Polymer Sci.* **1985**, *263*, 756.
4. Hoffmann, H.; Oetter, G.; Schwandner, B. *Progr. Colloid & Polymer Sci.* **1987**, *73*, 95.
5. Chang, D.L.; Rosano, H.L.; Woodward, A.E. *Langmuir* **1985**, *1*, 669.
6. Imae, T.; Araki, H.; Ikeda, S. *Colloids and Surfaces* **1986**, *17*, 221.
7. Benjamin, L. *J. Phys. Chem.* **1964**, *68*, 3575.
8. Rathman, J.F.; Scamehorn, J.F. *Langmuir* **1988**, *4*, 474.
9. Chang, D.L.; Rosano, M.J.L. in *Structure/Performance Relationships in Surfactants*; Rosen, M.J., Ed.; ACS Symposium Series 253; American Chemical Society: Washington, D.C., 1984; p. 129.
10. Abe, M.; Kato, K.; Ogino, K. *J. Colloid Interface Sci.* **1989**, *127*, 328.
11. Rathman, J.F.; Christian, S.D. *Langmuir* **1990**, *6*, 391.
12. For a typical synthetic experimental procedure, see *Organic Synthesis*; Collective Vol. 4, John Wiley & Sons, Inc., 1963, p. 612.
13. Cameron, D. G.; Kauppinen, J. K.; Moffatt. D. J.; Mantsch, H. H. *Appl. Spec.* **1982**, 36, 245.
14. Goddard, E.D.; Kung, H.C. *J. Colloid Interface Sci.* **1968**, *27*, 247.
15. Wiley, R. H.; Slaymaker, S. C. *J. American Chem. Soc.* **1957**, 79, 2233.
16. Katritzky, A. R.; Beard, J. A. T.; Coats, N. A. *J. Chem. Soc.* **1959**, *16*, 3680.
17. Katritzky, A. R.; Hands, A. R. *J. Chem. Soc.* **1958**, *14*, 2195.
18. Mathis-Noël, R; Wolf, R.; Gallais, F *Compt. Rend. Acad. Sci. (Paris)* **1956**, 242, 1873.
19. Kawai, T.; Umemura, J.; Takenaka, T.; Kodama, M.; Ogawa, Y.; Seki, S. *Langmuir* **1986**, 2, 739.
20. Umemura, J.; Kawai, T.; Takenaka, T. *Mol. Cry. Liq. Cry.* **1984**, 112, 293.
21. Kawai, T.; Umemura, J.; Takenaka, T.; Kodama, M.; Seki, S. *J. Colloid Interface Sci.* **1985**, 103, 56.
22. Casal, H. L.; Mantsch, H. H. *Biochim. et Biophys. Acta* **1984**, 779, 384.
23. Cameron, D. G.; Umemura, J.; Wong, P. T. T.; Mantsch, H. H. *Colloids and Surfaces* **1982**, 4, 131.
24. DeRuig, W. G. *Appl. Spec.* **1977**, 31, 122.
25. Zerbi, G.; Conti, G.; Minoni, G.; Pison, S.; Bigotto, A. *J. Phys. Chem.* **1987**, 91, 2386.
26. Snyder, R. G. *J. Chem. Phys.* **1967**, 47, 1316.
27. Umemura, J.; Mantsch, H. H.; Cameron, D. G. *J. Colloid Interface Sci.* **1981**, 83, 558.
28. Dluhy, R. A.; Moffatt, D.; Cameron, D. G.; Mendelsohn, R.; Mantsch, H. H. *Can. J. Chem.* **1985**, 63, 1925.
29. Dluhy, R. A.; Mendelsohn, R.; Casal, H. L.; Mantsch, H. H. *Biochemistry*, **1983**, 22, 1170.
30. Maroncelli, M.; Qi, S. P.; Strauss, H. L.; Snyder, R. G. *J. American Chem. Soc.* **1982**, 104, 6237.
31. Scheuing, D.R.; Weers, J.G. *Langmuir* **1990**, *6*, 665.

RECEIVED August 10, 1990

INTERFACIAL PHENOMENA

Chapter 8

Fourier Transform Infrared Spectroscopy of Langmuir–Blodgett and Self-Assembled Films

An Overview

Abraham Ulman

Corporate Research Laboratories, Eastman Kodak Company, Rochester, NY 14650–2109

This overview describes the different FTIR techniques used for the study of Langmuir-Blodgett (LB) and self-assembled (SA) films. It also reviews some of the recent literature in an attempt to give the reader a more complete picture on the area.

Infrared (IR) spectroscopy is our everyday tool for the study of molecular packing and orientation in ultrathin organic films. We use FTIR in two main modes: attenuated total reflection (ATR) on a silicon, germanium, or ZnSe crystals, with s-polarized (parallel, ‖, electrical field parallel to the x-axis in Figure 1) and p-polarized light (perpendicular, ⊥, electrical field parallel to the y-axis in Figure 1) to evaluate surface coverage and molecular orientation, and grazing-angle (GA) to study monolayers and films on metallic surfaces, and to study the orientation of specific transition dipoles in the monolayer. In this overview we describe the experimental techniques, and mention some of the contributions to this area.

Attenuated Total Reflection (ATR) Spectroscopy

A typical set-up for a FTIR ATR experiment is described in Figure 1. A Brewster angle rotating Ge polarizer is place in the IR beam, in front of the C face of an ATR crystal, which can be made of Si, Ge, or ZnSe. A separate background spectra are recorded for the s- and p-polarization, and the crystal is used as a substrate either for LB or SA monolayer samples.

0097–6156/91/0447–0144$06.00/0

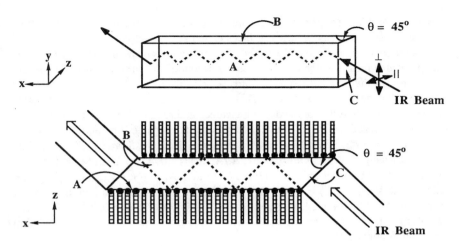

Figure 1. A schematic description of an optical set-up for ATR measurements.

The cross section area of the *collinear* IR beam is ~1 cm² and thus sufficient to cover the entire area of the C face of the ATR crystal, and as a result, a complete coverage of faces A by the IR radiation is achieved. Hence, the measured absorbance should be proportional to the fraction of the total area of faces A and C covered by the monolayer. Of course, all six faces of the crystal are covered with the monolayer film. However, only the A faces contribute to the measured signal *via* internal reflection. This is because the area of the C faces is only ~ 7% of the total area (A + C) in a typical ATR crystal, and the differences between transition mode (in the C faces) and ATR mode (in the A faces) are not very large. Therefore, it was suggested by Maoz and Sagiv that no corrections for this effect are needed (1).

The theory for analysis of ATR data was developed by Harrick (2,3), and applied by Haller and Rice (4). The following relations were predicted by Maoz and Sagiv for the ATR experiment described in Figure 1 (1).

(*a*.) For a perfect all-*trans* molecular orientation perpendicular to the surface with all vibrations parallel to the surface of the substrate (e.g., -CH₂- stretching vibrations, see Figure 5), the following equation can be written:

$$D = \frac{A_s}{A_p} = \frac{0.42d + 5d_e \bot_y \cos 45^\circ}{0.42d + 5d_e \| _x \cos 45^\circ} \qquad (1)$$

where D is the dichroic ratio, and is defined as the ratio between the absorbance recorded with perpendicular (A_s), and parallel (A_p) polarizations, respectively (note that parallel and perpendicular polarizations are defined in respect to the xz plane in Figure 1), d is the thickness of the film, $d_e\bot_y$ is the effective thickness of the film for the perpendicular polarization (electrical field parallel to the y axis in Figure 5), $d_e\|_x$ is

the effective thickness of the film for the x component of the parallel polarization, 0.42 and 5 are the areas (in cm^2) of faces C and A, respectively (for a 50 x 10 x 3 mm ATR crystal), and cos 45° corrects the expression given be Harrick (2) for the sampling area factor, which is now explicitly accounted for in equation 1. The reader is referred to the original paper of Harrick and du Pré for detailed discussion on effective thickness of thin films (3). This equation takes into account the contribution at faces A, as well as that of the direct transmission through faces C (for which $A_p = A_s$). If we insert the refractive indices for the crystal, organic film, and N_2 ($n_1(Ge) = 4.03$, $n_1(Si) = 3.42$, $n_1(ZnSe) = 2.42$, n_2(organic film) = 1.5, and $n_3(N_2) = 1$) into equation 1 we get: $A_s/A_p(Ge) = 1.061$, $A_s/A_p(Si) = 1.094$, $A_s/A_p(ZnSe) = 1.245$. If we now omit the contribution from faces C we get: $A_s/A_p(Ge) = 1.071$, $A_s/A_p(Si) = 1.103$, and $A_s/A_p(ZnSe) = 1.260$. For absorbance measured with nonpolarized IR radiation ($A = 1/2A_s + 1/2A_p$) we get: A(Ge) / A(Si) /A (ZnSe) = 0.598 / 0.704 / 1.
(*b*.) For random molecular orientation where vibrations are randomly distributed in space the following equation (2) can be written:

$$D = \frac{0.42d + 5d_e \perp_y \cos 45^\circ}{0.42d + 5d_e || \cos 45^\circ} \qquad (2)$$

where all notations are as above and $d_e|| = d_e||_x + d_e||_z$. Using the same refractive indices we get: $A_s/A_p(Ge) = 0.879$, $A_s/A_p(Si) = 0.897$, and $A_s/A_p(ZnSe) = 0.969$, and A(Ge) / A(Si) / A(ZnSe) = 0.584 / 0.690 / 1.
(*c*.) For perfect molecular orientation with all vibrations perpendicular to the surface $A_s = 0$ and $A_p \propto d_e||_z$. Vibrations that are perpendicular to the surface are more difficult to detect by ATR (usually we use grazing angle spectroscopy for this purpose). This is because of the fact that in the ATR mode $d_e||_z < d_e||_x$. For example, the A_p values predicted for transitions perpendicular to the surface are smaller than those predicted for parallel transitions of same intrinsic intensity (transition moment dipoles) by factors of 2.22 (Ge), 2.10 (Si), and 1.67 (ZnSe). Therefore, it is recommended to use ZnSe when possible for analysis of molecules with both parallel and perpendicular transitions, and where grazing angle spectroscopy is not available.

Most of the monolayer and multilayer films that are usually investigated fall within the limits of cases *a* and *b*. Vibrations belonging to case *c* are, for example, the symmetric stretching of the carboxylate group, or the symmetric stretching of the methyl group (see Figure 5) in well-oriented monolayers. When going from case *a* to case *c* without changing the number of molecules *per* unit film area we get the following relations:

$$A_{s(oriented)}/A_{s(random)} = 1.50 \text{ for all materials}$$
$$A_{p(oriented)}/A_{p(random)} = 1.223 \text{ (Ge), } 1.211 \text{ (Si), } 1.153 \text{ (ZnSe)}$$
$$A_{(oriented)}/A_{(random)} = 1.352 \text{ (Ge), } 1.346 \text{ (Si), } 1.324 \text{ (ZnSe)}$$

These differences in the magnitude of the measured absorbances are due to the redistribution of the transition dipoles in space, from a situation where all are in the x,y plane (e.g., $v_s(CH_2)$, $v_a(CH_2)$ in an all-*trans* perpendicular orientation, case *a*, to a uniform distribution of the transition dipoles along the x,y, and z directions, as in case *c*.

It is important to emphasize that the above discussion does not take into account the optical anisotropy of the oriented films (5), and assuming the same refractive index for all the sample. However, if the films have the same molecular density, such as, for the first approximation, complete monolayers (5-7), these constraints are not critical. On the other hand, these considerations may be important in partial monolayers, since significant deviations may arise when going down in surface concentration. The refractive index of the film, n_2, is a monotonic function of the molecular density, varying from 1.00, in the limit of zero surface coverage, to about 1.50, in a complete closely-packed monolayer.

The effective film thickness parameters, $d_e\perp_y$ and $d_e\|_x$ are proportional to n_2, while $d_e\|_z$ is proportional to $1/n_2^3$ (2). Therefore, measured absorbances in submonolayer coverage may not be expected to vary linearly with the surface coverage. The reader is referred to reference 1 for more detailed discussion.

In order to evaluate quantitatively the orientation of vibrational modes from the dichroic ratio in molecular films, we assume a uniaxial distribution of transition dipole moments in respect to the surface normal. (z-axis in Figure 1). This assumption is reasonable for a crystalline-like, regularly ordered monolayer assembly. An alternative, although more complex model is to assume uniaxial symmetry of transition dipole moments about the molecular axis, which itself is tilted (and uniaxially symmetric) with respect to the z-axis. As monolayers become more liquid-like, this may become a progressively more valid model (8,9). We define ϕ as the angle between the transition dipole moment M and the surface normal (note that $0^\circ \leq \phi \leq 90^\circ$). The absorbance due to the components E_x, E_y, and E_z of the electric field of the evenescent wave (2,10), in the ATR experiment are given by equations 3 - 5 (8).

$$A_z = \frac{1}{2}M^2E_z^2\sin^2\phi \qquad (3)$$

$$A_x = \frac{1}{2}M^2E_x^2\sin^2\phi \qquad (4)$$

$$A_y = \frac{1}{2}M^2E_y^2\sin^2\phi \qquad (5)$$

Light polarized parallel to the incident plane (p-polarized) has components in the x and z directions, while light polarized perpendicular to the incident plane (s-polarized) has only a y-component. Therefore, the dichroic ratio is given by equation 6.

$$D = \frac{A_s}{A_p} = \frac{A_y}{A_x + A_z} \qquad (6)$$

This equation is written in terms of A_s and A_p (defined above), omitting the contribution of the entrance and exit faces (C in Figure 1). The observed dichroic ratio D_{obsd} should, in principle, include a contribution A_T from the transmission absorbance from the entrance and exit faces (equation 7).

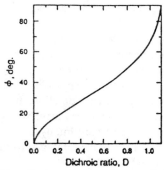

Figure 2. Plot of orientation angle φ vs dichroic ratio D.

$$D_{obsd} = \frac{A_s + A_T}{A_p + A_T} \qquad (7)$$

It was suggested above to neglect the contribution A_T on the basis of the small area of the entrance and exit faces (1). We examined this suggestion in a transmission IR experiment with a monolayer of molecule 1 on Si (see in part 3) (11). This experiment gave only a very weak spectrum, with an absorbance at 2920 cm^{-1} of only 6% of the absorbance in the ATR mode (using p-polarized light), which would affect D by less than 0.01 at values close to D = 1.0, and is, therefore, in agreement with setting $D \approx D_{obsd}$.

The expressions for the electric field components E_x, E_y, and E_z have been evaluated by Haller and Rice (4), based on a theory developed by Harrick (2). We insert these expressions into equation 3 - 5, and 6, calculate a plot of φ vs D (Figure 2), and estimate the unknown angle φ for a given monolayer from this plot.

Kopp et al. studied the rearrangement of LB layers of tripalmitin using ATR FTIR spectroscopy (12). They reported that these LB layers undergo a spontaneous transition from a liquid crystalline state to a microcrystalline one, and that this transition is accompanied by a shift of the CH_2 bending band from 1469 cm^{-1} to 1473 cm^{-1}, and of the ester group motion from 1169 cm^{-1} to 1182 cm^{-1}. Ohnishi et al. studied mono- and multilayers of cadmium arachidate on glass, and reported a regular perpendicular alignment of the alkyl chains (13). Hartstein et al. reported that thin metal overlayers or underlayers enhanced the absorption from monolayers by a factor of 20 (14). The total enhancement, including contributions from the ATR geometry, was almost 10^4. Okamura et al. studied monolayers of 1,2-dipalmitoyl-3-*sn*-phosphatidyl-choline (DPPC) on a germanium plate (15). They reported that the alkyl chains in these layers were oriented vertically to the Ge surface, with all-*trans* configuration, irrespective of the surface pressure on the film transfer. This result suggested that throughout the surface pressure examined, there were islands or surface micelles on the surface. Kimura, et al. studied LB films of stearic acid with 1-9 monolayers deposited on Ge (16). They examined the CH_2 scissoring band and suggested that the alkyl chains of the stearic acid in the first monolayer are in a hexagonal or pseudohexagonal subcell packing, and that they are perpendicular to the surface, and therefore are free to rotate around their molecular axis. On the other hand, in LB films thicker than two monolayers, alkyl chains in monolayers other than the first one crystallized in a monoclinic form and were packed alternately, with a tilt of ~30° with respect to the surface normal. Maoz and Sagiv used ATR FTIR to investigate oxidation reaction of double bonds with penetrated of permanganate ions (MnO_4^-) in self-assembled monolayers (17). Davies and Yarwood studied ATR FTIR of LB films of ω-tricosenoic acid on SiO_2/Si substrates (18). They reported that the integrated intensity increases roughly linearly with increasing thickness, and that there was a degree of random variation in the alkyl chain tilt. Kamata et al. studied the enhancement in IR ATR spectra of LB films of stearic acid overcoated with gold and silver films, and reported that, for the same metal thickness, gold is more effective than silver (19). In another paper, the same authors expanded the study to other metals and reported the order Au > Ag > Pt ≈ Cu > Ni > Al for the enhancement of the asymmetric CH_3 stretching in one stearic acid LB layer overcoated samples (20). They explain the origin of the enhancement order in terms of the complex dielectric constant of the metal, and its oxidation potential. Recently, Naselli et al. studied LB films (6-12 monolayers) of semifluorinated fatty acid $[C_8F_{17}(CH_2)_{10}CO_2H]$ (21). They report that in this case, unlike its hydrocarbon analogue, the chains are inclined due to packing constraints dictated by the considerably larger cross section of the fluorocarbon chain. In their paper they bring a detailed discussion on the IR. We bring these numbers here as a reference for LB and SA films of fluorocarbon amphiphiles (the polarization in parantheses indicates the direction with respect to helix axis): 1300 cm^{-1} ($\perp$), v_a(CC), δ(CCC); 1242 cm^{-1} ($\perp$), v(CF_2), r(CF_2); 1213 cm^{-1} (||), δ(CCC), v(CC); 1153 cm^{-1} ($\perp$), v_s(CF_2), δ(CF_2); 638 cm^{-1} (||), ω(CF_2), r(CF_2); 553 cm^{-1} ($\perp$), δ(CF_2); 503 cm^{-1} (||), ω(CF_2); (Note the designations: v = stretching, δ = deformation, r = rocking, and ω = wagging).

Grazing-Angle Spectroscopy

Reflection-absorption (RA) or grazing-angle spectroscopy, is a very useful technique that gives information about the direction of transition dipoles in a sample. Figure 3 present an optical setup for a grazing-angle experiment (22).

Theoretical consideration of the IR spectroscopy of monolayers adsorbed on a metal surface showed that the reflection-absorption spectrum is measured most efficiently at high angles of incidence, and that *only parallel* component of incident light gives measurable absorption species (23). Figure 4 presents a schematic description of a monomolecular film on a mirror, with the incident light and direction of the polarization. Figure 5 presents, in detail, an alkyl thiol molecule on a metal surface. Note the direction of the different transition dipoles. Thus, while both the symmetric and asymmetric methylene vibrations (v_s and v_a, respectively), are parallel to the metal plane, the symmetric vibration and both asymmetric vibrations of the methyl group have components that are perpendicular to the surface. Therefore, the

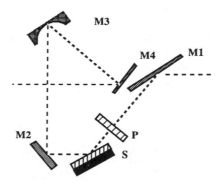

Figure 3. An optical arrangement for measuring polarized reflection-absorption spectra of a monolayer or thin film on the surface of a mirror S at near grazing incidence. P is a polarizer, M1, M2, M3 and M4 are mirrors.

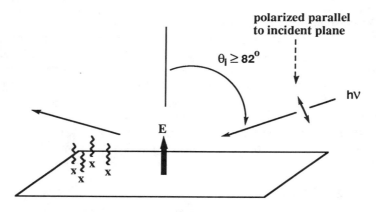

Figure 4. A schematic diagram of the grazing-angle IR experiments.

methylene groups in a perpendicular, all *trans* alkyl chain will not be picked up by the p-polarized light. Of course, once the alkyl chain tilts from the normal to the plane, the symmetric and asymmetric vibrations of the methylene groups are no longer parallel to the surface and thus will appear in the grazing-angle spectra. The intensity of the methylene vibrations in the spectra is a direct function of the tilt angle of the alkyl chain. This unique property of the grazing-angle experiment enables the calculation of molecular orientation from FTIR spectra.

A modification of the simple grazing-angle experiment is the multiple reflection - absorption spectroscopy, which was first described by Gaines (Figure 6) (24). It has a unique value for the detection of order-disorder transformations in the structure of a

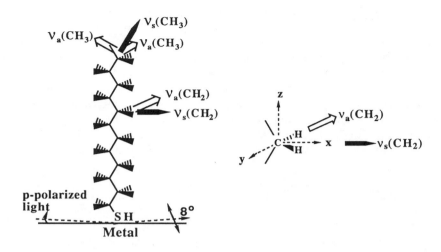

Figure 5. An alkyl thiol on a metal surface in a grazing-angle FTIR experiment (left), and an in-plane diagram of the CH_2 group and its transition dipoles (right). Note that the CH_2-plane is parallel to that of the substrate, and that both $v_s(CH_2)$ and $v_a(CH_2)$ are in that plane, where $v_s(CH_2)$ and $v_a(CH_2)$ are orthogonal to each other.

monolayer, resulted from its interaction with an external physical or chemical agent. Gaines *et al.* used two parallel Al/glass mirrors, separated by a 3 mm-thick, frame-shaped, spacer, and mounted it in the plate holder of an ATR attachment, set at 75^0 angle of incident. At this geometry most of the incident beam radiation is used in the measurement. They were able to further enhance the signal by using only the parallel polarization (∥). The multiple reflections makes this technique more sensitive than the regular grazing angle, which is a single-reflection experiment.

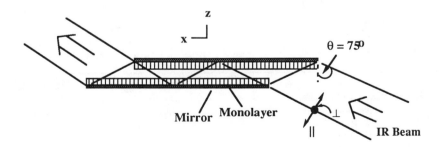

Figure 6. An optical set-up for the study of monolayers in a reflection-absorption mode.

One of the early experiments in multiple-reflection grazing-incidence was carried out by Francis and Ellison (25). However, the pioneeric work in the area of grazing-angle FTIR spectroscopy in monolayers is that of Allara and Swalen, who related peak positions and intensities to film structure at the molecular level (26). The characterization of alkyl thiol monolayers on gold by grazing-angle FTIR was made by Porter et al. (27). They carried out a very detailed study of the structure as a function of the chain length. A part of Table I in their paper is presented in Table I. Figure 7 presents the C-H stretching region of the IR spectrum of a self-assembled monolayer of $CH_3(CH_2)_{21}SH$ on gold with the peak assignments (28-32). The assignment of the bands in the C-H stretching region is important since these peaks appear in any IR spectra regardless of the assembling technique. The band at 2965 cm^{-1} is the asymmetric in-plane C-H stretching mode of the CH_3 group, $v_a(CH_3,ip)$. The bands at 2973 and 2879 cm^{-1} are the symmetric C-H stretching mode of the CH_3 group, $v_s(CH_3,FR)$. The abbreviation FR denotes the Fermi resonance interaction, which causes the splitting of this band (30). The bands at 2917 and 2850 are the asymmetric, v_a, and symmetric, v_s, C-H stretching of the CH_2. Porter et al. used $v_a(CH_3)$ and $v_s(CH_2)$ for structural interpretation simply because there is a minimal overlap between them. A careful inspection of the data in Table I and Figure 7, suggests that: (*a.*) the formation of the two-dimensional assembly does not alter the peak positions as compared to the bulk spectra. (*b.*) both peak positions and intensities are the function of the alkyl chain length (n).

Table I. Peak Positions for $CH_3(CH_2)_nSH$ C-H Stretching Modes in Crystalline and Liquid States and Adsorbed on Gold

Structural group	Stretching mode	crystalline[b]	liquid[c]	Peak positions[a]			
				n = 21	n = 15	n = 9	n = 5
-CH$_2$-	v_a	2918	2924	2918	2918	2920	2921
	v_s	2851	2855	2850	2850	2851	2852
CH$_3$-	$v_a(ip)$	d	d	2965	2965	2966	2966
	$v_a(op)$	2956	2957	e	e	e	e
	$v_s(FR)$	f	f	2937	2938	2938	2939
	$v_s(FR)$	f	f	2879	2879	2878	2878

a. ± 1 cm^{-1}. *b.* Crystalline-state positions determined for $CH_3(CH_2)_{21}SH$ in KBr. *c.* Liquid-state positions determined for $CH_3(CH_2)_7SH$. *d.* The $v_a(ip)$ is masked by the strong $v_a(op)$ in the crystalline- and liquid-state spectra. *e.* The position of $v_a(op)$ could not be determined because of low signal-to-noise ratio. This is the result of the orientation of this mode with respect to the surface. *f.* Both $v_s(Fr)$ bands are masked by the strong $v_a(CH_2)$ band.
SOURCE: Ref. 27. Copyright 1987 The American Chemical Society.

It is useful, at this point, to refer to the work of Snyder et al. (28-29). The trend toward higher peak frequencies as the length of the alkyl chain decreases in the

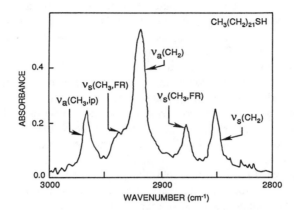

Figure 7. The C-H region in the IR spectrum of $CH_3(CH_2)_{21}SH$ on gold (Reprinted with permission from ref. 27. Copyright 1987 The American Chemical Society.)

monolayer is in agreement with their findings for polyethylene. Thus, they reported 2920 cm^{-1} for the $v_a(CH_2)$ mode in crystalline polyethylene, which is 8 cm^{-1} lower than the peak position in the liquid state (2928 cm^{-1}). Also reported was the value of 2850 cm^{-1} for $v_s(CH_2)$ in crystalline polyethylene, which is 6 cm^{-1} lower than the peak value in the liquid state (2856 cm^{-1}). The same trend in peak position is reported in Table I. for the solid and liquid phases, where changing from the solid to the liquid phase shifts the above modes by +6 cm^{-1} and +4 cm^{-1}, respectively. Let us compare the frequencies for the monolayers with alkyl chains ($n > 15$). Here, $v_a(CH_2) = 2918$ cm^{-1}, the same frequency as in the KBr spectrum of the solid $CH_3(CH_2)_{21}SH$. On the other hand, this mode appears at 2921 cm-1 in the $CH_3(CH_2)_5SH$ ($n = 5$) monolayer approaching 2924 cm^{-1}, which is the value in the pure liquid ($n = 7$). Thus, it is recommended that the FTIR spectrum of a monolayer will be compared with the spectrum of the pure material in solution (or neat for a liquid) and in KBr (for a solid) to establish the range of frequencies for the different modes. This is especially true where the molecule exhibits a richer spectrum due to different functionalities. Such an FTIR study may help to establish, for example, that there are different degrees of packing (e.g., liquid-like and solid-like) in different parts of the monolayer, due to different types of interactions between the different functional groups.

Golden et al. studied the grazing-angle spectra of LB and self-assembled arachidate monolayers on aluminum surfaces (32). Rabolt et al. studied cadmium arachidate monolayers on AgBr for transmission, and on Ag for grazing-angle experiments. They found that, independent of the substrate, the chains of the fatty acid salt are oriented within a few degrees of the normal to the surface of the substrate (33). Bonnerot et al. studied LB layers of docosanoic and ω-tricosenoic acids on aluminum surfaces (34). They observed a structural transition when the thickness increased from 1 to > 7 layers. The hexagonal subcell became orthorhombic, and the axis of the chain, perpendicular to the substrate for the first layer, tilted progressively to reach a final limit of 23° for behenic acid, and 18° for ω-tricosenoic acid. Naselli et al. used grazing-angle FTIR spectroscopy to study thermally induced order-disorder transition in LB films of cadmium arachidate (35). They suggested a pretransitional disordering, prior to melting. The introduction of aromaticity into the molecule (e.g., tetradecylbenzoic acid), improved the high temperature stability of the films, probably due to enhanced intermolecular interactions between the head groups. Hallmark et al. compared self-assembled monolayers of octadecyltrichloro-silane (OTS), and LB monolayers of cadmium arachidate, and concluded that the chain axes in the OTS monolayer has some degree of tilt, while that in the arachidate salt monolayer are oriented normal to the substrate (36). Dote and Mowery studied the orientation of LB monolayers of stearic acid and perdeuterated stearic acid [$CD_3(CD_2)_{16}COOH$] on gold and native-oxide aluminum (Al_2O_3/Al) surfaces (37). This is an important work since it was observed that monolayers on aluminum aged with time and realigned to a more perpendicular orientation. In another paper the authors assign bands in deuterated alkyl chains (37). The $v_a(CD_2)$ and $v_s(CD_2)$ appear at 2194 and 2086 cm^{-1}, respectively, while the $v_a(CD_3)$ appears at 2212 and 2221 cm^{-1} for the in-plane and out-of-plane, respectively, and at 2103 and 2076 cm^{-1} appear $v_s(CD_2)$, and $v_s(CD_3)$, respectively.

Recent Results from Our Laboratory

We have used FTIR-ATR extensively to study SA monolayers (e.g., **1**), and multilayers of alkyltrichlorosilanes (11, 39, 40). Figure 8a presents the bulk (CCl_4 solution) IR spectrum of **1**, while Figures 8b, and 8c, present the grazing angle and ATR spectra of a monolayer of **1** on aluminized silicon, and silicon ATR prism, respectively.

$$CH_3(CH_2)_8\text{—}\langle\!\!\bigcirc\!\!\rangle\text{—}O\text{—}(CH_2)_{11}\text{-}SiCl_3$$

1

In the bulk spectrum (Figure 8a) of **1**, the methylene stretching vibrations ($v_a(CH_2)$ at ~2920 cm^{-1} and $v_s(CH_2)$ at ~2855 cm^{-1}) are extremely intense and bury the methyl vibrations ($v_a(CH_3)$ at ~2965 cm^{-1} and $v_s(CH_3)$ at 2880 cm^{-1}). In the grazing-angle spectra, by contrast, the methylene vibrations are greatly diminished in relative intensity. For perpendicular alkyl chains, both the $v_a(CH_2)$ and the $v_s(CH_2)$ will be oriented parallel to the substrate surface; consequently, their interaction with the electric field vector should approach zero.

We suggested, as a first approximation to assume that the transition dipole moments of both the methyl and methylene vibrations are equal on a per hydrogen basis, and *estimated* the orientation of the chains in monolayers of **1**, and also of OTS by consideration of the measured $v_s(CH_2)/v_s(CH_3)$ intensity ratio.

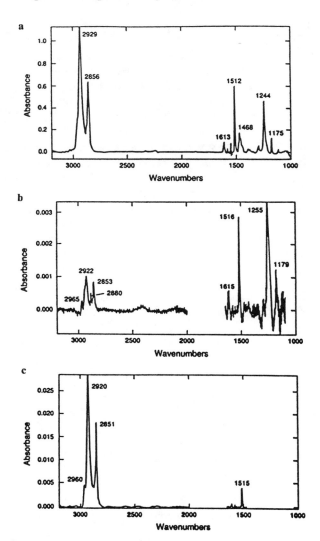

Figure 8. (a) FTIR spectrum of **1** in CCl$_4$, (b) C-H stretching region of the grazing angle external specular reflection IR spectrum of a monolayer of **1** on an aluminized silicon wafer, and (c) as in (b) for a monolayer of OTS on an aluminized silicon wafer. Spectra are recorded at 76° incidence, 1000 scans, 2 cm^{-1} resolution, and the baselines have been adjusted to zero absorbance.

These results suggested alkyl chain tilts of ~15° for OTS and ~26° for **1**. This "tilt" is the angle between the chain axis, which bisects the C-C bonds, and the surface normal.

The ATR-IR results for the alkyl chain axis orientation of 1 and OTS are qualitatively similar to what we found using grazing-angle spectroscopy. The IR results for the monolayers of 1 are consistent and suggest that the alkyl chains are more tilted than observed for OTS. This is likely due to the steric bulk of the phenoxy group, which increases the chain-chain spacing. Consequently, the chains tilt in order to maintain van der Waals contact. We estimated a ~20° tilt for the aromatic ring in 1, which was somewhat surprising, since 35° would be predicted from a model postulating perpendicular alkyl chains and normal C-C-C bond angles of 109.5° (Figure 9) This, however, can be rationalized by comparing the cross-sectional area of the alkyl chain and the phenyl ring. Thus, while the cross-sectional area of an all-*trans* alkyl chain is ~20 Å^2 (40), simple calculations show that the cross-sectional area of a phenyl ring is from 21.8 to 25.2 Å^2 (up to 25% larger, depending on the phenyl orientation in the monolayer). At ~20° tilt the cross-sectional area of the phenyl ring is smaller than at 35°. Thus, in order to achieve more complete close-packing the molecules tilt and the benzene rings are "squeezed" into a more perpendicular orientation.

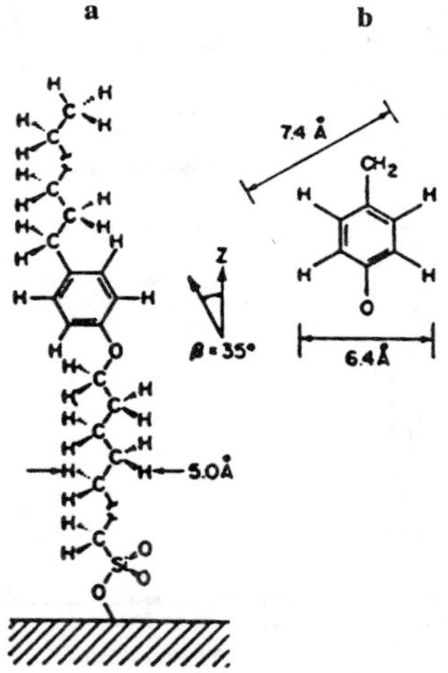

Figure 9. (a) Representation of 1 in a monolayer with fully extended alkyl chains that are perpendicular to the surface. The length of the alkyl chains has been shortened for clarity. The C_1-C_4 axis of the phenyl ring has a 35° tilt (β) from the substrate surface normal (z-axis). (b) Dimensions of the phenyl ring, calculated from the usual bond lengths and van der Waals radii.

We have also used grazing-angle FTIR to study alkanethiol monolayers on gold and silver surfaces. Figures 10 and 11 present the spectra in the C-H stretch region of 2800-3100 cm^{-1} for monolayers of octadecanethiol on gold (ODT/Au) and on

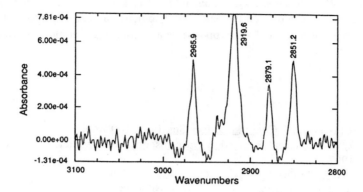

Figure 10. Grazing-angle FTIR spectrum of ODT/Au.

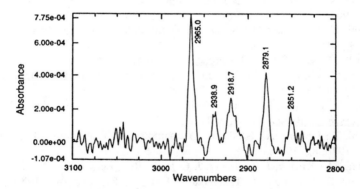

Figure 11. Grazing angle FTIR spectrum of ODT/Ag.

silver (ODT/Ag) surfaces. We measured intensity ratios $v_s(CH_2)/v_s(CH_3)$ (at ~ 2851/2880 cm^{-1}), and $v_a(CH_2)/v_a(CH_3)$ (at ~ 2920/2965 cm^{-1}) of 1.25, and 1.47 for ODT/Au, respectively. For a monolayer of octadecanethiol on silver (ODT/Ag) the intensity ratios were 0.44 and 0.45, respectively. The $v_a(CH_2)$ and $v_s(CH_2)$ peak positions for ODT/Ag agree well with those of condensed-phase alkanes (42), as well as with those of $CH_3(CH_2)_{21}SH$ in KBr ($v_a(CH_2)$ = 2918 cm^{-1}, $v_s(CH_2)$ = 2851 cm^{-1}) (27). These intensity ratios suggest that the alkyl chains in the monolayers on silver are much less tilted than the corresponding monolayers on gold. Furthermore, the spectrum of the ODT/Ag monolayer resembles, for example, the spectrum of adsorbed cadmium arachidate on an oxidized aluminum surface (31). However, whereas in this system the $v_a(CH_2)/v_a(CH_3)$ was ~1, in the ODT/Ag monolayer this ratio is only 0.45. Thus, the alkyl chains in ODT/Ag are more perpendicular than those in cadmium arachidate on the oxidized aluminum surface. It is interesting to note that a tilt angle of $0 \pm 15°$ was proposed for a multilayer of cadmium arachidate on a silicon surface (43).

References

1. Maoz, R; Sagiv, J. J. Colloid Interface Sci. 1984, 100, 465.
2. Harrick, N. J. Internal Reflection Spectroscopy; Wiley-Interscience: New York, 1967.
3. Harick, N. J.; du Pre, F. K. Appl. Opt. 1966, 5, 1739.
4. Haller, G. L.; Rice, R. W. J. Phys. Chem. 1970, 74, 4386.
5. Den Engelsen, D. J. Opt. Soc. Am. 1970, 61, 1460.
6. Blodgett, K. B.; Langmuir, I. Phys. Rev. 1937, 51, 964.
7. Tomar, M. S. J. Phys. Chem. 1974, 78, 947.
8. Zbinden, R. Infrared Spectroscopy of High Polymers; Academic Press: New York, 1964.
9. Fringeli, U. P.; Schadt M.; Rihak, P.; Gunthardt, Js. H. Z. Naturforsch. A 1974, 31, 1098.
10. Mirabella, F. M., Jr. Appl. Spectrosc. Rev. 1985, 21, 45.
11. Tillman, N.; Ulman, A.; Schildkraut, J. S.; Penner, T. L. J. Am. Chem. Soc. 1988, 110, 6136.
12. Kopp, F.; Fringeli, U. P.; Muehlethaler, K.; Guenthard, Hs. H. Z. Naturforsch. C: Biosci. 1975, 30, 711.
13. Ohnishi, T.; Ishitani, A.; Ishida, H.; Yamamoto, N.; Tsubomura, H. J. Phys. Chem. 1978, 82, 1989.
14. Hartstein, A.; Kirtley, J. R.; Tsang, J. C. Phys. Rev. Lett. 1980, 45, 201.
15. Okamura, E.; Umemura, J.; Takenaka, T. Biochim. Biophys. Acta 1985, 812, 139.
16. Kimura, F.; Umemura, J.; Takenaka, T. Langmuir 1986, 2, 96.
17. Maoz, R.; Sagiv, J. Langmuir 1987, 3, 1034.
18. Davies, G. H.; Yarwood, J. Spectrochim. Acta Part A 1987, 43, 1619.
19. Kamata, T.; Umemura, J.; Takenaka, T. Bull. Inst. Chem. Res., Kyoto Univ. 1987, 65, 170.
20. Kamara, T.; Umemura, J.; Takenaka, T. Bull. Inst. Chem. Res., Kyoto Univ. 1987, 65, 179.
21. Naselli, C.; Swalen, J. D.; Rabolt, J. F. J. Chem. Phys. 1989, 90, 3855.
22. Ishitani, A; Ishida, H.; Soeda, F.; Nagasawa, Y. Anal. Chem. 1982, 54, 682.
23. Blanke, J. F.; Vincent, S. E.; Overend, J. Spectrochim. Acta, Part A 1976, 32, 163.
24. Gains, G. L., Jr. Insoluble Monolayers at Liquid-Gas Interfaces; Interscience: New York, 1966.

25. Francis, S. A.; Ellison, A. H. J. Opt. Soc. Am. 1959, 49, 131.
26. Allara, D. L.; Swalen, J. D. J. Phys. Chem. 1982, 86, 2700.
27. Porter, M. D.; Bright T. B.; Allara, D. L.; Chidsey, C. F. D. J. Am. Chem. Soc. 1987, 109, 3559.
28. Snyder, R. G.; Strauss, H.; Ellinger, C. A. J. Phys. Chem. 1982, 86, 5145.
29. Snyder, R. G.; Maroncelli, M. Strauss, H. L.; Hallmark, V. M. J. Phys. Chem. 1986, 90, 5623.
30. Hill, I. R.; Levin, I. W. J. Chem. Phys. 1979, 70, 842.
31. Allara, D. L.; Nuzzo, R. G. Langmuir 1985, 1, 52.
32. Golden, W. G.; Snyder, C. D.; Smith, B. J. Phys. Chem. 1982, 86, 4675.
33. Rabolt, J. F.; Burns, F. C.; Schlotter, N. E.; Swalen, J. D. J. Electron. Spectrosc. Relat. Phenom. 1983, 30, 29.
34. Bonnerot, A. Chollet, P. A.; Frisby, H.; Hoclet, M. Chem. Phys. 1985, 97, 365.
35. Naselli, C.; Rabe, J. P.; Rabolt, J. F.; Swalen, J. D. Thin Solid Films 1985, 134, 173.
36. Hallmark, V. M.; Leone, A.; Chiang, S. ; Swalene, J. D.; Rabolt, J. F. Polym. Prepr. (Am. Chem. Soc. Div. Polym. Chrm.) 1987, 28, 22.
37. Dote, J. L.; Mowery, R. L. J. Phys. Chem. 1988, 92, 1571.
38. Mowery, R. L.; Dote, J. L. Mikrochim. Acta 1988, 2, 69.
39. Tillman, N.; Ulman, A.; Penner, T. L. Langmuir 1989, 5, 101.
40. Tillman, N.; Ulman, A.; Elman, J. F. Langmuir, 1989, 5, 1020.
41. Kuhn, H.; Mobius, D. In Physical Methods in Chemistry, A. Weissberger, B. Rossiter, Eds.; Wiley: New York, 1972; Part 3B, Vol. 1, p 577.
42. Snyder, R. G.; Hsu, S. L.; Krimm, S. Spectrochim. Acta Part A 1978, 34, 395.
43. Rabe, J. P.; Swalen, J. D.; Outka, D. A.; Stöhr, J. Thin Solid Films 1988, 159, 275.

RECEIVED July 26, 1990

Chapter 9

n-Alkanoic Acid Self-Assembled Monolayers

Adsorption Kinetics

S. H. Chen and C. W. Frank

Department of Chemical Engineering, Stanford
University, Stanford, CA 94305-5025

FTIR methods were used to study the kinetics of formation of self-assembled monolayers of n-alkanoic acids by adsorption from solutions. With reference to Langmuir-Blodgett deposited monolayers, the adsorption of n-alkanoic acids from hexadecane solutions onto glass and aluminum substrates was shown to lead to monolayer formation. Stearic acid adsorption, as monitored by FTIR, was found to follow a transient Langmuir adsorption model. Additionally, fluorescence measurements using a pyrene-tagged n-alkanoic acid probe co-adsorbed into the monolayers were used to investigate the dependence on chain length. An increase in the negative free energy of adsorption with increasing acid chain length was shown.

Organized organic molecular monolayers formed by spontaneous adsorption from solution, known as self-assembled (SA) films, have attracted a great deal of research interest in recent years(1-5) because of their potential technological and scientific applications(6,7). The formation of compact, well-organized structures that closely resemble Langmuir-Blodgett monolayers has been evidenced by studies employing a number of physical measurements. In the formation of ordered structures from amphiphilic molecules, e.g., in the Langmuir-Blodgett transfer process(7,8) as well as in the self-assembling of amphiphilic molecules(1-4), the head group-substrate interaction and the tail group-tail group interaction are the major contributing factors. Surfactants solvated in hydrophobic, nonpolar solvents adsorb onto a hydrophilic polar surface such that the hydrophilic head groups attach to the surface while the hydrophobic tails line up with one another, forming a hydrophobic exterior surface that is analogous to the solution phase. In such systems the van der Waals attractive interaction among the hydrophobic portions of the adsorbate molecules contributes to their alignment. The ability of these molecules to form organized monolayers increases with increasing length of the aliphatic chain(3,8). Clearly, both the substrate-head group binding and the van der Waals attraction among the hydrocarbon chains will control the adsorption kinetics. Allara, Nuzzo, Whitesides and coworkers(3,4) and Sagiv and coworkers(1,2), in their series of extensive studies using a number of characterization techniques, have demonstrated that SA films of structure similar to that of the Langmuir-Blodgett films can be produced under appropriate conditions. They also showed that the kinetics of adsorption is important in the formation of these organized two-dimensional films. In

0097-6156/91/0447-0160$06.00/0
© 1991 American Chemical Society

general, the well-known Langmuir isotherm is applicable for monolayer adsorption(9), but very little kinetic data have been reported for these types of systems.

Fourier-transformed infrared spectroscopy (FTIR), either in the transmission mode(10), the grazing incidence reflection (GI) mode(1,3) or the attenuated total reflection (ATR) mode(1,2), has been the most widely used experimental tool for the characterization and structure determination of SA monolayers. GI-IR is especially useful in determining the molecular orientation in the film structures because it senses only the vibrational component perpendicular to the substrate surface(1,3). Polarized ATR-IR can also be used to study molecular orientation(1,11). McKeigue and Gulari(12) have used ATR-IR to quantitatively study the adsorption of the surfactant Aerosol-OT.

Besides FTIR methods, other spectroscopic techniques such as fluorescence and UV spectroscopy may be considered as complementary experimental tools to FTIR. Fluorescence spectroscopy of systems containing small fractions of covalently bound fluorescent probes within the compound of interest can provide molecular-level information. The usefulness of the fluorescence probe approach has been demonstrated for Langmuir-Blodgett films(13).

In this work we utilized FTIR methods to examine the SA monolayers on flat, polar solid surfaces prepared from nonpolar solutions. We used ATR and GI FTIR measurements to characterize the material and bonding of the SA monolayers, and used transmission and ATR FTIR to monitor the dynamics of the SA adsorption process. With reference to measurements on standard Langmuir-Blodgett monolayer samples, we were able to quantify the SA kinetic results. We also used fluorescence spectroscopy of incorporated pyrene probes in SA mixed monolayer films as a simple method for the determination of the relative adsorption and thermodynamic constants.

Experimental

Sample Preparation. The homologous series of the even n-alkanoic acids, abbreviated as C_{10} through C_{22} were used as the adsorbates. Hexadecane (HD), which is nonpolar and has a rather high boiling point, was used as the solvent. Microscope glass slides and evaporated aluminum (on silicon wafers) were used as the substrates. Pyrene end-tagged hexadecanoic acid (Py-C_{16}) was used as the fluorescence probe.

SA monolayers were prepared by immersing the substrates in the n-alkanoic acid/HD solutions for a predetermined period of time. After removal from the solutions, any remaining liquid droplets on the surface or the edges of the substrates were blown off with a nitrogen jet. For the equilibrium monolayer characterization, relatively high solution concentration ($>10^{-3}$M) were used. For time-dependent kinetic studies, stearic acid (C_{18}) solutions of concentrations between 10^{-2} and 10^{-6}M were used. Immersion times from a few seconds up to 24 hours were used.

The reference samples of Langmuir-Blodgett monolayers of C_{16}-C_{28} monolayers were prepared using a Joyce-Loebl Langmuir trough. Film materials were deposited using solutions in CHCl$_3$. For the cadmium salt monolayers, a subphase of 2.5×10^{-4}M CdCl$_2$ solution was used. All monolayers were deposited at 30 dynes/cm and 15°C.

Pyrene-labeled SA mixed monolayers were prepared by adsorption from solutions of the desired concentration of a particular fatty acid, along with a small fraction (1-5%) of the probe Py-C_{16}. All solutions used were of total acid concentration of 5×10^{-3}M.

Sample Characterization. Infrared spectra of the adsorbed films were obtained with a Perkin-Elmer 1710 FTIR Spectrometer, equipped with a DTGS detector and a nitrogen-purged sample chamber. The transmission IR spectra (high frequency range; >2000 cm^{-1}) of the adsorbed species were directly measured on the glass slides. The

ATR-IR spectra were taken on the same spectrometer, using a multiple-internal-reflection attachment obtained from Harrick Scientific Inc, using an ATR crystal (one-pass parallelepiped KRS-5; 45°, 50×10×3mm) also from Harrick. The ATR-IR spectra of adsorbed species were obtained by pressing two film-covered substrates against the internal reflection surfaces of the ATR crystal. All spectra were obtained with 4 cm^{-1} resolution. In practice, it was usually necessary to average 1000-2000 scans in order to obtain spectra of acceptable signal-to-noise ratio. The reference (background) spectra were taken with clean substrates prior to adsorption.

The auxiliary film characterization tools used include contact angle goniometry and ellipsometry. A Ramé-Hart goniometer was used to measure the advancing contact angles of HD and water on substrates before and after adsorption of the fatty acids. A Gaertner ellipsometer with a He-Ne laser source was used to measure the film thickness. For the fluorescent probe studies, fluorescence emission spectra of the pyrene-doped monolayers were obtained with a Spex Fluorolog 212 spectrofluorometer. The excitation wavelength was set at 343 nm and the spectra were taken in the front-face mode.

Results

Characterization of Monolayers.

The change in surface wettability of the substrate was used as a qualitative indicator for monolayer adsorption. Clean glass and aluminum substrates were wetted by both water and HD. The contact angles were small (< 10°). After adsorption of the amphiphilic fatty acid molecules, the surface wettability was drastically reduced, as indicated by the sheeting-off of the solutions. The advancing water contact angle on equilibrated fatty acid/Al samples obtained was 105±2°, and the hexadecane contact angle was 47±2°. For fatty acid/glass samples, the hexadecane contact angle was close to that of fatty acid/Al. These values were similar to those previously reported(1,3). To confirm the formation of monolayers, the thickness of equilibrated fatty acid films C_{14}-C_{22}/Al, prepared from relatively concentrated (0.01M) solutions, were measured by ellipsometry. The results show that the film thickness correspond to the extended zig-zag molecular lengths of the film-forming molecules(3), suggesting that they are indeed monolayers with the chains organized approximately perpendicular to the substrates.

FTIR measurements were used to characterize the adsorbed monolayers. Transmission-IR spectra were obtained for the IR transparent range of glass (wavenumber>2000cm^{-1}), which contains the CH_2 and CH_3 stretch peaks of the hydrocarbon tails of the fatty acid molecules. A typical transmission-IR spectrum of two stearic acid monolayers (one monolayer on each side of a glass slide) adsorbed from 0.01M solution is shown in Figure 1(a). The symmetric and asymmetric CH_2 stretches (2850cm^{-1} and 2920cm^{-1}, respectively) and the symmetric and asymmetric CH_3 stretches (2890cm^{-1} and 2962cm^{-1}, respectively) were observed as expected.

ATR-IR measurements were used to obtain the full-range IR spectra of the adsorbed species. Two substrates with adsorbed films were pressed against the two ATR crystal surfaces so that the adsorbate was sampled by the internally reflected IR beam. A moderate pressure was necessary in order to "push" the substrate surface into the penetration range of the IR beam, which is on the order of several micrometers. Shown in Figures 1(b) and 1(c) are typical ATR-IR spectra of adsorbed stearic acid adsorbed at equilibrium on glass and Al substrates, respectively. The most significant difference between these two spectra is that only the 1420/1470cm^{-1} and 1580cm^{-1} peak, assigned to the symmetric and asymmetric COO$^-$ stretches were observed for C_{18}/Al, while the 1730cm^{-1} peak, assigned to the free acid C=O stretch, was seen in addition to the carboxylate stretches for C_{18}/glass. This implies that on the Al substrate the anchoring acid head groups were totally deprotonated into salt (COO$^-$)(3), while on glass only a portion of them were deprotonated on adsorption, the remainder remaining as free acid. The glass slides contain several metal oxides such as Na_2O, CaO, MgO, and Al_2O_3,

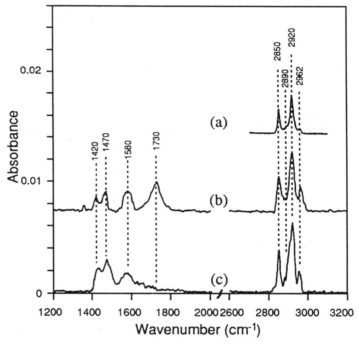

Figure 1.   FTIR spectra of C_{18} monolayers: (a) High-frequency range transmission-IR spectrum of two C_{18}/glass monolayers (one monolayer on each side), (b) ATR-IR spectra of C_{18}/glass, and (c) C_{18}/aluminum (with oxide layer on top); prepared from 0.01M solution in HD, immersion time=30min.

along with the major component SiO_2. Apparently these metal atoms present on the surface enhance the anchoring of the acid head groups, perhaps by chemical bonding, to the metal sites (salt formation), while only physisorption (polar interaction with free acid) occurs on the silicon sites.

GI-FTIR spectra for have also obtained for C_{18}/Al monolayers. The results were very similar to those of Allara and Nuzzo(4). The spectra show enhanced symmetric and asymmetric CH_3 stretches and diminished symmetric and asymmetric CH_2 stretches, as expected for ordered hydrocarbon chains approximately perpendicular to the substrate surface. Similar to the ATR-IR result, carboxylate peak was also observed in the GI-IR spectra. With these results, the formation of ordered monolayer structure is also supported.

Quantification of FTIR Measurements. The transmission-IR and ATR-IR absorbance can be used to quantitatively determine the amount of adsorbate present. To serve as a standard reference, Langmuir-Blodgett deposited C_{16}-C_{28} Cd^{+2} salt monolayers were used. The CH_2 stretch peak absorbances were correlated with the absolute surface concentration of CH_2 groups, based on measured deposition ratio and molecular area, together with the number of CH_2 units in the molecule. With monolayers deposited at 30 dynes/cm, the deposition ratios were generally close to 1 (1.0-1.2), except for the very long C_{28} (0.6-0.7). At this surface pressure, the molecular area of the acids were 18-20 Å^2/molecule, decreasing with increasing chain length. Figure 2 shows such a correlation between the transmission FTIR 2920 cm^{-1} peak absorbance and N_{CH_2}, the surface concentration of CH_2 groups. The Absorbance/CH_2 thus obtained can then be used to quantify the similar measurements for the SA monolayers. The values obtained for transmission-IR and ATR-IR are listed in Table I.

Table I. Reference Values of 2920 cm^{-1} Absorbance/CH_2

Type of Measurement	Absorbance/Å^2·CH_2 group
transmission-IR, glass	0.00425
ATR-IR, Al	0.00777

Transient Adsorption Behavior. The adsorption of C_{18} on glass or Al at various times of immersion from solutions of various adsorbate concentrations was followed by transmission-IR and ATR-IR. Contact angle measurement was also used to follow the surface wettability change for these samples. Both the IR absorption peak intensities and the contact angles increased initially with increasing time of immersion as well as increasing adsorbate concentration, and asymptotically reached plateau values at long immersion times and high solution concentrations. The time required to reach the plateau maxima increased with decreasing adsorbate concentration.

Figures 3(a) and 3(b) show the 2920cm^{-1} transmission IR peak intensity and the HD contact angle variation for C_{18}/glass, and Figures 4(a) and 4(b) the 2920cm^{-1} ATR-IR peak intensity and the water and HD contact angle variation for C_{18}/Al, respectively. Using the reference Absorbance/CH_2·Å^2 values in Table I, the surface concentrations of adsorbed C_{18} were also calculated and are shown in the right-hand scales of Figures 4(a) and 5(a). The plateau value of approximately 8.5×10^{-10} mol/cm^2 (corresponding to 19.6Å^2/molecule) agrees quite well with the well-known packing densities in L-B monolayers of simple amphiphilic compounds. The intensity of the peaks other than the one at 2920cm^{-1} also increased proportionately. The plateau intensities of the high

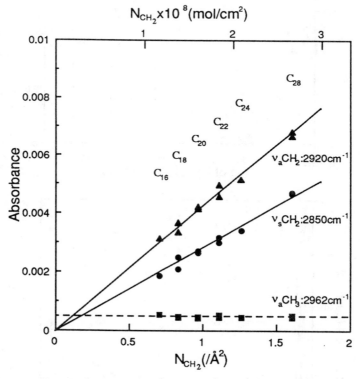

Figure 2. Correlation of 2920cm^{-1}, 2850cm^{-1}, and 2962cm^{-1} transmission-IR peak absorbances with surface concentration of -CH$_2$- units of L-B monolayers of C$_{16}$-C$_{28}$.

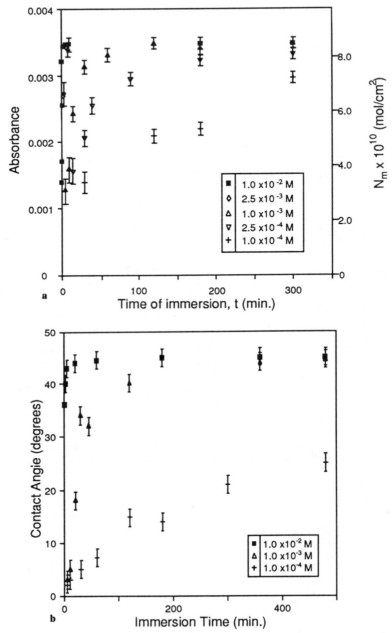

Figure 3.   Transient adsorption of C_{18} on glass monitored by (a) 2920cm^{-1}
transmission IR peak absorbance and (b) HD contact angle.
Surface concentration of C_{18} is also shown in (a).
(Adapted from ref. 5)

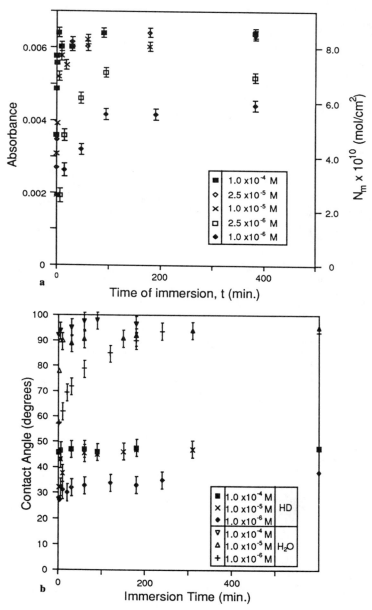

Figure 4.   Transient adsorption of C_{18} on (oxidized) aluminum monitored by
(a) 2920cm^{-1} ATR-IR peak absorbance and (b) HD contact angle.
Surface concentration of C_{18} is also shown in (a).
(Adapted from ref. 5)

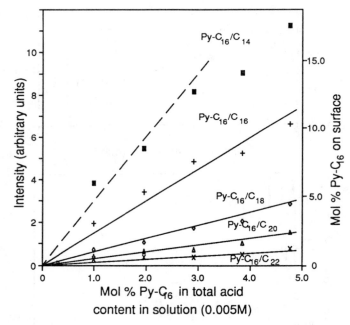

Figure 5. Fluorescence emission intensity of Py-C$_{16}$/C$_{xx}$ monolayers on aluminum from solutions of 0.005M total acid concentration. Surface molar percentage of Py-C$_{16}$ is also shown. (Adapted from ref. 5)

frequency peaks (CH_2 and CH_3 stretches) were quite reproducible, up to about ±5%, which is essentially the error of IR measurement at this intensity level. However, the reproducibility of the carbonyl and carboxylate stretch peaks in the lower frequency range were somewhat worse (±20%).

Fluorescence Of Monolayers Containing Pyrene-Labeled Probes. A fluorescence probe method was also used as a complementary technique to study the thermodynamics of SA film formation. Mixed monolayers containing the fluorescence probe pyrene hexadecanoic acid, Py-C_{16}, in host fatty acids of different lengths were prepared by adsorption from solutions containing mostly the host fatty acid and a small fraction of Py-C_{16} (approximately 1 to 5 mol %). All monolayers were prepared under equilibrium adsorption conditions. For fluorescence measurements only Al substrate was used because when glass is used an impurity fluorescence from glass interferes with the pyrene fluorescence.

The pyrene monomer emission intensity is related to the amount of Py-C_{16} adsorbed on the substrate. With host fatty acids of carbon number greater than about 14, the pyrene monomer emission intensity was found to vary with the preparation conditions. Figure 5 shows the 377 nm peak intensity of mixed monolayers prepared at equilibrium from solutions of 1 to 4.8% tagged acids in 0.005M total concentration. The intensity increased approximately linearly with the molar fraction of Py-C_{16} in solution, and decreased by about half as the carbon number of the host fatty acid increased by two. The fluorescence emission intensities were also normalized using similar measurements with Langmuir-Blodgett deposited Py-C_{16}/C_{20} (free acid) mixed monolayers containing 0.5-5% Py-C_{16}. The surface Py-C_{16} molar percentages thus determined are shown in the right-hand scale of Figure 5.

Discussion

Adsorption Kinetics -- Transient Langmuir Model. The parallelism of the increase in the contact angles and the FTIR absorbance suggests that these two measurements monitor the same quantity -- the surface coverage of the adsorbate. As has been shown by the normalized FTIR absorbance, the plateau maximum surface concentrations obtained indeed correspond to closely-packed monolayers. If we assume that the plateau adsorption at the highest solution concentrations corresponds to the full coverage of the surface "sites" of the substrate, as for the monolayer adsorption, the fractional coverages of the substrate surface can then be calculated from the normalized FTIR peak absorbances. From theoretical considerations, monolayer adsorption can be treated as a surface site-filling procedure, with the adsorption and desorption steps counteracting each other.

In deriving governing equations for this type of transient Langmuir adsorption kinetics, the diffusional mass transfer resistance has been neglected, as has previously been validated by Grow and Shaeiwitz(14) for adsorption of several surfactants from aqueous solutions. In general, for the adsorption of surfactants onto planar surfaces, the surface concentration of the adsorbate is so small that the adsorption step is always rate-controlling. In this manner, the material balance equation can be written as(14,15):

$$\frac{d\theta}{dt} = \frac{k_a}{N_0} \cdot c \cdot (1 - \theta) - \frac{k_d}{N_0} \cdot \theta \qquad (1)$$

where θ is the fractional surface coverage calculated from the normalized IR peak intensity, t is the adsorption time, k_a and k_d are the adsorption and desorption rate constants, N_0 is the surface adsorbate concentration at full coverage, and c is the solution concentration of adsorbate.

Upon integration with the initial condition $\theta=0$ at t=0 one obtains

$$\theta = \frac{k_a \cdot c}{k_a \cdot c + k_d}\left\{1 - \exp\left[-\frac{k_a}{N_0}\left(c + \frac{k_d}{k_a}\right)\cdot t\right]\right\} \qquad (2)$$

This equation reduces to the Langmuir isotherm at equilibrium, i.e., as $t \to \infty$

$$\theta_{eq} = \frac{k_a \cdot c}{k_a \cdot c + k_d} = \frac{c}{c + \kappa} \qquad (3)$$

where $\kappa \equiv k_d/k_a \propto \exp(\Delta G_a^o/RT)$ and ΔG_a° is the free energy of adsorption at infinite dilution.

For the adsorption of C_{18} from solutions of concentrations between 10^{-2} and 10^{-4}M onto glass slides, the transient absorbance data previously shown can be used for kinetic calculations. Figure 6(a) shows the transient surface coverage obtained by normalizing the 2920cm^{-1} IR absorption peak intensities, such as those shown in Figure 3(a), to the plateau maximum value, which corresponds to the closely-packed monolayer. For solutions of concentration on the order of 10^{-2}M, full coverage ($\theta=1$) was reached in a couple of minutes. On the other hand, for very dilute solutions, it took hours to reach the equilibrium coverages θ_{eq}, which were less than unity. The equilibrated coverage data, obtained with glass slides immersed in solutions for 24 hours, were reasonably well fitted with equation 3, as shown in Figure 6(b). Using the value of κ obtained, the functional form of equation 2 was used to fit the data (as shown by the curves in Figure 6(a)) with an exponential constant, which should equal $[(-k_a/N_0)(c+k_d/k_a)]$. As a check for fitting, the exponential constants obtained were plotted against c, as shown in Figure 6(c). As can be seen, the fitting of the model was reasonably good. The numerical values of the kinetic constants, obtained from the data-fitting together with the surface concentration at full coverage $N_0=8.5\times10^{-10}$ mol/cm^2 previously determined, are given in Table II.

Figures 7(a)-(c) show similar results for the adsorption of stearic acid from solutions of concentrations between 10^{-4} and 10^{-6}M onto aluminum substrates, again obtained by monitoring the normalized 2920cm^{-1} IR absorption peak intensity. The trends observed in this system are the same as those of the C_{18}/glass case, but the range of solution concentration over which significant coverage variation can be seen is quite different. For solution concentrations greater than about 0.001M, the adsorption was very rapid, much faster than in the C_{18}/glass case. The numerical values of the kinetic constants are also given in Table II.

Table II. Kinetic Constants

System	κ (mol/cm^3)	ΔG_a° (Kcal/mol)	k_a (cm/s)	k_d (mol/cm^2·s)
C_{18}/glass	$(1.5\pm0.1)\times10^{-8}$	-7.3 ± 0.1	$(1.0\pm0.1)\times10^{-5}$	$(1.5\pm0.1)\times10^{-13}$
C_{18}/Al	$(5.0\pm0.4)\times10^{-10}$	-9.2 ± 0.1	$(5.2\pm0.4)\times10^{-4}$	$(2.6\pm0.3)\times10^{-11}$

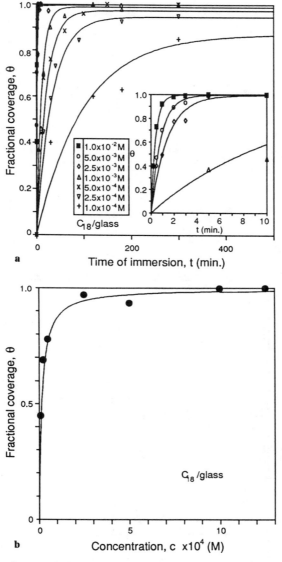

Figure 6.   Langmuir kinetic model fitting of C_{18} adsorption on glass: (a) fractional coverage as a function of time and solution concentration, (b) equilibrium coverage as a function of concentration, and (c) check of fitting of equation 2 by the correlation of the fitted exponential constant and concentration.

(Reprinted with permission from ref. 5. Copyright 1989 American Chemical Society.)

Continued on next page.

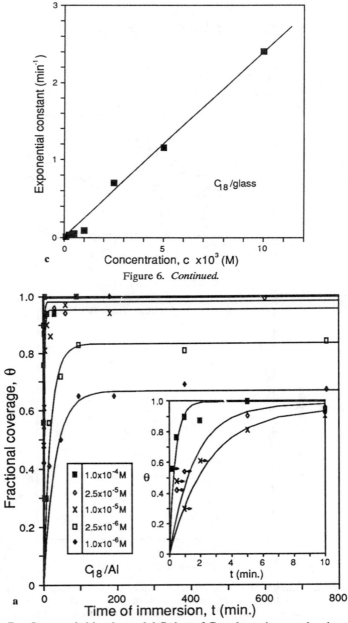

Figure 6. *Continued.*

Figure 7.   Langmuir kinetic model fitting of C_{18} adsorption on aluminum:
(a) fractional coverage as a function of time and solution
concentration, (b) equilibrium coverage as a function of
concentration, and (c) check of fitting of equation 2 by the
correlation of the fitted exponential constant and concentration.

Continued on next page.

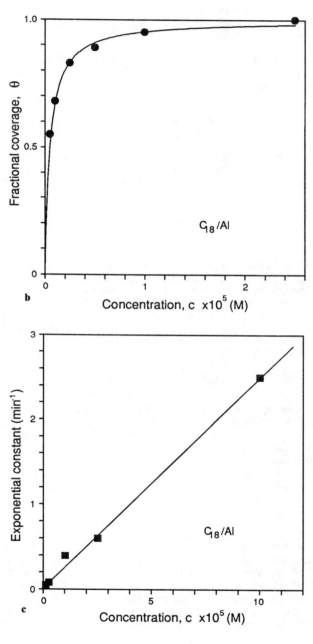

Figure 7. *Continued.*

Compared with literature data for adsorption of surfactants from aqueous solutions on oxide surfaces[14,15], the kinetic data obtained in this work for C_{18}/glass are of similar orders of magnitude to the former systems. The values of $1/\kappa$ and k_a for C_{18}/Al are greater than those for oxide adsorbent studied, indicating the strong adsorbing ability of metallic aluminum (even oxidized) relative to the mineral oxides.

Relative Adsorption Kinetic Constants From Fluorescence Data. Using a two-component, competitive adsorption Langmuir isotherm, θ_g, the fractional coverage of the guest molecule can be expressed as

$$\theta_g = \frac{c_g/\kappa_g}{c_g/\kappa_g + c_h/\kappa_h + 1} \propto I \qquad (4)$$

where κ_g and κ_h are the equilibrium constants, as defined in equation 3, for the guest molecule Py-C_{16} and the host fatty acid respectively, and I is the monomer fluorescence (377nm) intensity. Under the experimental conditions used, $c_g \ll c_h$, $\theta_h \approx 1$, and equation 4 reduces to

$$\theta_g = \frac{c_g}{c_g + \kappa_g/\kappa_h} \qquad (5)$$

Fitting the data in Figure 5 gives the ratios of the constants κ_h between host fatty acids (such as $\kappa_h(C_{22})/\kappa_h(C_{20})$, $\kappa_h(C_{20})/\kappa_h(C_{18})$, etc...). The ratios, in turn, give the difference of the free energy of adsorption between various host fatty acids. Using the value for C_{18}/Al obtained previously as the starting point, the $-\Delta G_a°$ values for other fatty acids can then be calculated. For fatty acids long enough to achieve the Langmuir adsorption behavior, based on the energy additivity, the contribution made by the head group, $-\Delta G_a^{o,h}$, and that of the N_c methylene groups to the free energy of adsorption of a surfactant may be separated as

$$-\Delta G_a^o = -\Delta G_a^{o,h} + N_c \cdot W \qquad (6)$$

where W is the energetic contribution of the carbon chain per unit CH_2 group. As shown in Figure 8, there is a monotonic increase in the negative free energy of adsorption of 230 ± 20 cal/mol per -CH_2- in the fatty acid tail, which is the contribution of the van der Waals interaction among the aliphatic chains to the stabilization of the adsorbed amphiphilic molecules. This result agrees with literature values obtained by Bigelow et.al.[16] using wettability measurements for the adsorption of surfactants from oil/melt phases to polar surfaces.

Conclusions

FTIR methods were used to study self-assembled n-alkanoic acid monolayers on glass and Al substrates. The composition, structure, and the kinetics of formation of the SA monolayers were investigated. With reference to Langmuir-Blodgett deposited monolayers, FTIR measurements showed that the SA fatty acid monolayers are closely-packed with adsorbate surface concentrations of about 8.5×10^{-10} mol/cm^2. Using FTIR measurements, we demonstrated the transient Langmuir adsorption kinetics and obtained the kinetic constants for C_{18}/glass and C_{18}/Al systems. With additional fluorescence methods, we also studied the variation of the free energy of adsorption of

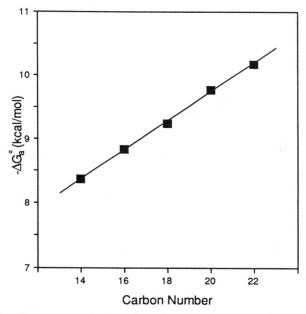

Figure 8.   Free energy of adsorption of *n*-alkanoic acids on (oxidized) aluminum at infinite dilution as a function of chain length.

(Reprinted with permission from ref. 5. Copyright 1989 American Chemical Society.)

the *n*-alkanoic acid molecules with their chain length. An increase in $-\Delta G_a^\circ$ with increasing chain length was observed and attributed to the dispersive interaction energy contribution to the self-assembly process.

Acknowledgments

We acknowledge the helpful discussions with Prof. R. F. W. Pease. We also appreciate the financial support of the Office of Naval Research (through contract N00014-87-K-0426).

Literature Cited

1. Maoz, R.; Sagiv, J. *J. Colloid Interface Sci.*, **1984**, *100*, 465.
2. Maoz, R.; Sagiv, J. *Langmuir*, **1987**, *3*, 1034.
3. Allara, D. L.; Nuzzo, R. G. *Langmuir*, **1985**, *1*, 45. *Langmuir*, **1985**, *1*, 52.
4. Troughton, E. B.; Bain, C. D., Whitesides; G. M., Nuzzo, R. G.; Allara, D. L.; Porter, M. D. *Langmuir*, **1988**, *4*, 365.
5. Chen, S. H.; Frank, C. W. *Langmuir*, **1989**, *5*, 978.
6. Swalen, J. D.; Allara, D. L.; Andrade, J. D.; Chandross, E. A.; Garoff, S.; Israelachvili, J.; McCarthy, T. J.; Murray, R.; Pease, R. F. W.; Rabolt, J. F. Wynne, K. J.; Yu, H. *Langmuir*, **1987**, *3*, 932.
7. Roberts, G. G. *Adv. Phys.*, **1985**, *34*, 475.
8. Gaines, G. L. *Insoluble Monolayers at Liquid-Gas Interfaces*, Interscience: New York, NY, 1966.
9. Tadros, Th. F. Ed.; *Surfactants*, Academic Press: New York, NY, 1984.
10. Hasegawa, M.; Low, M. J. D. *J. Colloid Interface Sci.*, **1969**, *29*, 593.
11. Yang, R. T.; Low, M. J. D.; Haller, G. L.; Fenn, J. *J. Colloid Interface Sci.*, **1973**, *44*, 249.
12. McKeigue, K.; Gulari, E. In *Surfactants in Solution*; Mittal, K. L.; Lindman, B. Eds.; Plenum Press: New York, NY, 1984; Vol. 2, p 1271.
13. Murakata, T.; Miyashita, T.; Matsuda, M. *Langmuir*, **1986**, *2*, 786.
14. Grow, D. T.; Shaeiwitz, J. A. *J. Colloid Interface Sci.*, **1982**, *86*, 239.
15. Trogus, F. J.; Sophany, T.; Wade, W. H. *Soc. Pet. Eng. J.*, **1977**, *17*, 337.
16. Bigelow, W. C.; Glass, E.; Zisman, W. A. *J. Colloid Sci.*, **1946**, *2*, 563.

RECEIVED September 17, 1990

Chapter 10

Polymeric Langmuir–Blodgett Films
Fourier Transform Infrared Spectroscopic and Gas Transfer Studies

Pieter Stroeve[1,3], Greg J. Spooner[2,3], Paul J. Bruinsma[1,3], Lawrence B. Coleman[2,3], Christian H. Erdelen[4], and Helmut Ringsdorf[4]

[1]Department of Chemical Engineering, [2]Department of Physics, and [3]Organized Research Program on Polymeric Ultrathin Films, University of California, Davis, CA 95616
[4]Institut für Organische Chemie, Universität Mainz, D–6500 Mainz, Federal Republic of Germany

We report here on the structure and gas transport properties of asymmetric membranes created by the Langmuir-Blodgett deposition of ultra-thin polymeric lipid films on porous supports. Transmission and grazing angle FTIR spectroscopy provide a measure of the level of molecular order in the n-alkyl side-chains of the polymeric lipid. The level of orientational order was monitored as a function of the temperature. Gas permeation studies as a function of membrane temperature are correlated to the FTIR results.

Langmuir-Blodgett (LB) multilayers are formed by the dynamic transfer of mono-molecular layers of amphiphilic molecules (the Langmuir layer) from the air-subphase interface to a solid support (1). It is now possible to design and fabricate macromolecular assemblies constructed of sequentially deposited monomolecular layers (2,3). The LB method is capable of producing systems from one to many hundreds of layers, each layer one molecule thick. The deposition of each monolayer is effected by the surface pressure of the Langmuir layer, the transfer (or dipping) speed, the ambient and subphase temperatures, the substrate surface, and the composition of the subphase. LB films and multilayer structures have potential applications in a wide variety of fields, including guided wave optical devices, non-linear optical devices, chemical sensors, permselective membranes, resists used in semiconductor processing, and optical computing (2,3,4). We are investigating the use of LB films in the fabrication of asymmetric membranes by depositing polymeric LB films on a porous substrates. Such membranes may have a potential for efficient permselective gas separation because of high permeation rates and good selectivity. The fabrication of asymmetric membranes with polymeric LB multilayers was previously achieved by LB deposition of reactive monomer amphiphiles followed by

0097–6156/91/0447–0177$06.00/0

deposition of reactive monomer amphiphiles followed by polymerization on the rigid substrate(5). However, *in situ* polymerization often causes cracking of the LB film(6). Recently, preformed polymerized lipids were deposited on porous polypropylene (Celgard 2400) and polytetrafluroethylene (GORE-TEX) and the transport properties of these membranes studied (7,8). To completely cover the porous substrate, the LB monolayer must be robust enough to bridge membrane pores and other surface features and form a solid, pore free film. One polymer, a 2,terpoly 1:5:5 co-polymerized lipid, was found to seal the porous support and demonstrated selectivity of CO_2 and CH_4 over N_2 (7). In this earlier report both mass transfer measurements and Fourier transform infrared (FTIR) studies were carried out on several polymerized lipid membranes(8).

FTIR spectroscopy has become a standard technique for investigating the structure and level of orientational order of ultrathin films, that is, films in the submicron range (2.5 to 500 nm) (9). By combining transmission spectroscopy with infrared reflection-absorption spectroscopy (this technique is also referred to as grazing incidence reflection), the orientation of the functional groups of the molecules in the film can be investigated. These measurements are nondestructive to the film and can be conducted at a range of temperatures and pressures.

We report here on the structure and gas transport properties of asymmetric membranes produced by the LB deposition of a polymeric lipid on porous supports. The effects of temperature on the structure and gas transport is described. The selectivity of CO_2 over N_2 permeation through the LB polymer films is determined. The polymerized lipid used in this study contains tertiary amines which may influence the CO_2 selectivity over N_2. The long term objective of our work is to understand how structure and chemistry of ultrathin films influence the gas permeation.

Experimental Details

Materials and Film Preparation. The molecular structure of the polymerized lipid referred to as CO-1.5 is shown in Figure 1. The material is a co-polymer of a double 18-carbon alkyl chain lipid with a side-group spacer and five main-chain spacer groups. The purpose of the spacer chains is to allow more free volume for the lipid chains to orientationally order normal to the polymer backbone. The lipid chain contains an amide, and the main-chain spacer groups contain tertiary amines. The polymer was synthesized following the general procedures given by Laschewsky *et al.* (10).

For the preparation of Langmuir-Blodgett multilayers, a Joyce-Loebl Model 4 trough was used. The polymer was spread from a chloroform solution at a concentration of approximately 1 mg/ml. The de-ionized water subphase was passed through an activated carbon adsorber and then distilled in a glass still. Before deposition the

Langmuir layer was repeatedly compressed and expanded until reproducible, equilibrium pressure-area (Π-A) curves were obtained. Two different porous membrane supports were studied. GORE-TEX is an expanded polytetrafluoroethylene membrane whose microstructure is shown in Figure 2a. As can be seen in the scanning electron micrograph, the surface and presumably the bulk of the membrane consists of beads of polytetrafluoroethylene interconnected by thin strands of the fluoropolymer. The membranes used had a nominal thickness of 25 μm and an effective pore size of 0.02 μm. Although the nominal pore size of the GORE-TEX is 0.02 μm, the microphotograph shown in Figure 2a shows that on the surface there are openings which are considerably larger than 0.02 μm. The second support used was 25 μm thick Celgard 2400. Celgard is produced from polypropylene, and the topography of the uncoated Celgard membrane is shown in Figure 2b. The surface of the Celgard membrane contains slit-like pores which are approximately 40 nm wide and 400 nm long. Below the surface the pores connect in a network that extends to a similar porous surface on the opposite side of the membrane. As Figure 2b demonstrates, the Celgard surface does not have an uniform distribution of the slit-like pores. Some regions of the surface contain none or only a few of the pores.

The porous supports were soaked and rinsed with HPLC grade chloroform and dried before coating. Asymmetric membranes were fabricated by coating one side of the porous support with multiple LB layers. The polymer was deposited at 21°C with a surface pressure of 55 mN m^{-1} and a dipping speed of 0.25 cm/min. A waiting period of 30 minutes was used between the upstroke and downstroke. This surface pressure is higher and the dipping speed is lower than our previous study and these conditions lead to high quality sealed asymmetric membranes. The LB deposition of CO-1.5 was found to be Y-type. Figure 2c shows a typical LB deposited CO 1.5 film on a Celgard substrate. The scanning electron micrograph was obtained using the upper stage of an International Scientific Instruments (Model DS 130) scanning electron microscope with a secondary electron detector. To enhance LB polymer film contrast, samples were exposed over several days to vapors from a 1% osmium tetraoxide solution. A carbon coating was evaporated on the sample to provide charge dissipation.

The advantage of using an evaporated carbon coating over the sputtered gold coating is that carbon has a low atomic number and provides little contrast itself. The carbon coating is essentially transparent to the electron beam. The result is that the contrast in the scanning electron micrograph is due to the polymeric film itself rather than the conductive coating in the case of a sputtered gold coating.

The "bubble" like features seen in the micrograph may be due to damage of the film because of excessive heating of the film during the evaporation of carbon onto the sample. Micrographs obtained without

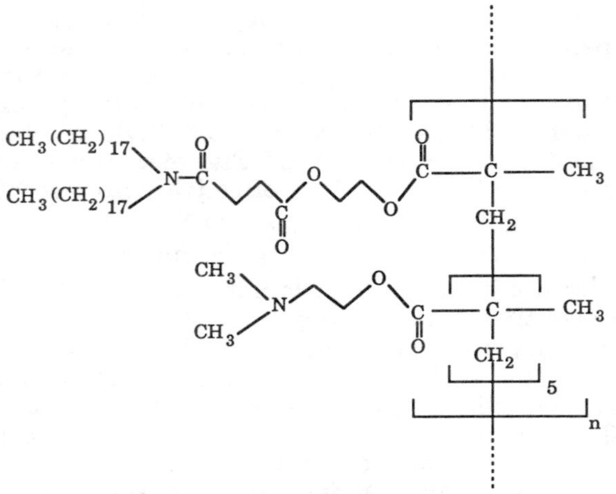

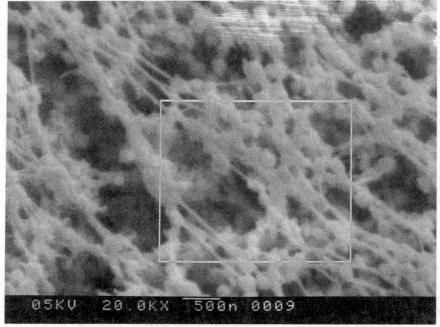

Fig. 1. Chemical structure of lipid co-polymer CO-1.5.

Fig. 2a. Scanning electron micrograph of bare surface of GORE-TEX with 0.02 μm effective pore size. Magnification 20,000 x. Bar represents 500 nm.

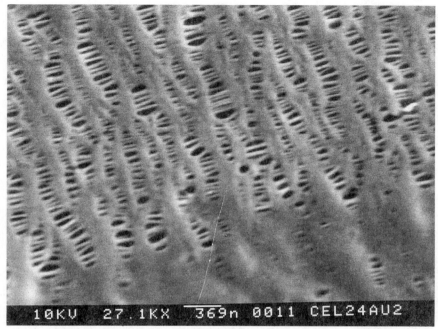

Fig. 2b. Scanning electron micrograph of bare surface of Celgard 2400. Magnification 27,100 x. Bar represents 369 nm.

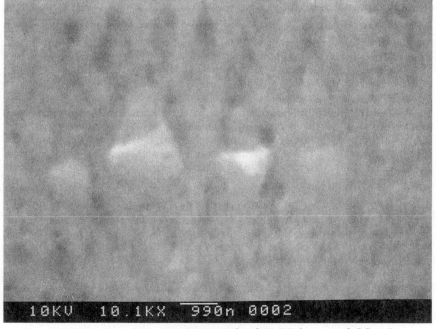

Fig. 2c. Scanning electron micrograph of 20 LB layers of CO-1.5 on Celgard 2400. Magnification 10,100 x. Bar represents 990 nm.

a conductive coating made by balancing charge buildup with accelerating voltage and sample tilt do not show such features, but are of poor quality. In the future a backscatter detector will be used to eliminate the need for a conductive coating.

Gas permeabilities were measured in a Skirrow-Barrer gas diffusion apparatus operating at a partial vacuum (7). The asymmetric membrane separated a large upstream gas chamber and a small downstream gas chamber. The gas pressure change in the downstream chamber was monitored with a differential pressure transducer (MKS Baratron 586/386).

The infrared spectra were obtained with a Nicolet 510 P spectrometer equipped with a room temperature DTGS detector. For infrared reflection-absorption spectroscopy (IRAS) studies a fixed angle (80°) Spectratech reflectance accessory was used with the LB films deposited on metallized glass slides. Spectra were taken over a range of 400 - 4000 cm^{-1} with a resolution of 2 cm^{-1} by co-adding 128 interferograms. Spectra were obtained at elevated temperatures using a temperature controlled sample holder. Room temperature spectra were also obtained of films annealed at elevated temperatures. Both of these sample modifications allowed us to study the temperature induced molecular reorganization of the LB films (8,11).

Results and Discussion

Infrared Spectroscopy. The GORE-TEX substrate was chosen for its complete fluorocarbon composition. As it contains no hydrocarbon, we are able to investigate the IR spectra of the polymerized lipid coating, free from substrate band interference. The GORE-TEX is opaque near 1200 cm^{-1} due to absorptions of the CF$_2$ and CF$_3$ groups. However in the hydrocarbon stretch region (2800 to 3000 cm^{-1}), the transmittance is featureless and near 50%. In our previous study (8), we reported on the FTIR spectra of LB films on GORE-TEX. In comparison to GORE-TEX, the Celgard membrane is hydrocarbon based and is therefore opaque in the hydrocarbon stretch region (2800 to 3000 cm^{-1}). In addition the Celgard transmission spectrum, as shown in Figure 3, is partially obscured by interference fringes. These interference features are the result of multiple reflections of the IR radiation between the front and rear surfaces of the Celgard membrane. The nonuniformity of the Celgard membrane porosity also results in a variation of the effective optical thickness. Because of the variation of the optical thickness between regions on a single Celgard film, as well as between individual pieces of Celgard material, the interference fringes cannot be removed by the usual spectral subtraction or division techniques.

For the FTIR investigations, GORE-TEX is an ideal substrate membrane on which to deposit the LB films. However, the GORE-TEX membrane is readily deformed which leads to cracks in the LB film during handling and mounting into the diffusion cell. This problem

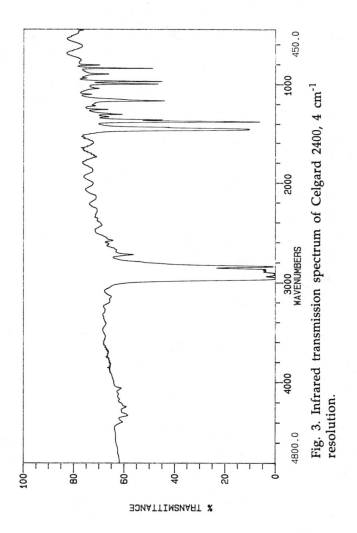

Fig. 3. Infrared transmission spectrum of Celgard 2400, 4 cm^{-1} resolution.

was not encountered using Celgard. However the strong 2800-3000 cm^{-1} absorptions and the interference fringes due to multiple reflections made FTIR transmission spectra of little value. To investigate the structure of LB films on the surface of Celgard we coated the Celgard membrane with 400Å of gold or aluminium, which then served as a reflecting substrate for IRAS measurements. The diffusion of gases was not affected as the metal layer did not bridge the pores. Comparing the IRAS spectra of LB films of CO-1.5 deposited on metallized Celgard with that deposited on metallized glass slides, we find the absolute and relative strengths of the CH$_2$ and CH$_3$ stretch bands to be unchanged. However the strength of the CH$_2$ wag bands (1267 cm^{-1} and 1242 cm^{-1}) and the C-O-C stretch (1157cm^{-1}) are greater for the films deposited on metallized glass substrates. This indicates some differences in the orientation and structure of the main-chain spacers as well as the side-group spacers which couple the alkyl chains to the backbone. We continue to investigate these substrate sensitive modifications. As our primary interest is in the orientational order and structure of the alkyl (lipid) chains, we continued our FTIR study of the CO-1.5 films using metallized glass slides for grazing angle absorption- reflectance measurements.

Figure 4 shows a representative IRAS spectra for a CO-1.5 film cast from a chloroform solution on an aluminized glass slide. The band frequencies and mode assignments are given in Table I. These band assignments have been made using an analysis of a similar polymer (11). In addition to the symmetric and asymmetric CH$_2$ stretches, two different C=O stretches are identified, one which is a component of an ester group (1736 cm^{-1}) and the other which is contained in a tertiary amide group(1650 cm^{-1}). A grazing angle reflectance spectrum of 20 LB layers of CO-1.5 at room temperature is shown in Figure 5. The similarities between Figures 4 and 5 indicates that the cast film has some partial ordering of the side chains. Additionally, there is a feature in the LB film spectra at 2770 cm^{-1} which does not appear in the cast film spectra. We tentatively assign this band to a nitrogen-methyl stretch (-N-CH$_3$). The absence of this feature in the cast film indicates that the N-CH$_3$ dipole lies in the plane of the substrate, which implies that the spacer groups also lie in the substrate plane. Additionally, the CH$_2$ stretch bands are more intense in the cast film spectra than the LB deposited film spectra, indicating that the CH$_2$ dipoles are oriented on the average closer to the normal in the cast film. Since these dipoles are nearly perpendicular to the chain axis, these spectra indicate that the alkyl tails in the cast film are significantly inclined from the normal relative to the orientation of the chains in the LB films.

Infrared reflection-absorption spectra of 42 molecular layers of CO-1.5 deposited on aluminized glass are shown in Figure 6. Spectra are presented at room temperature (~ 27 °C) and at the elevated

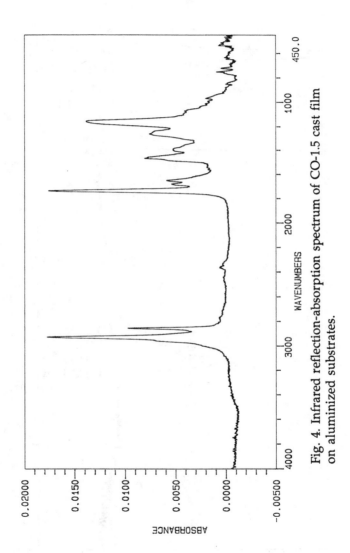

Fig. 4. Infrared reflection-absorption spectrum of CO-1.5 cast film on aluminized substrates.

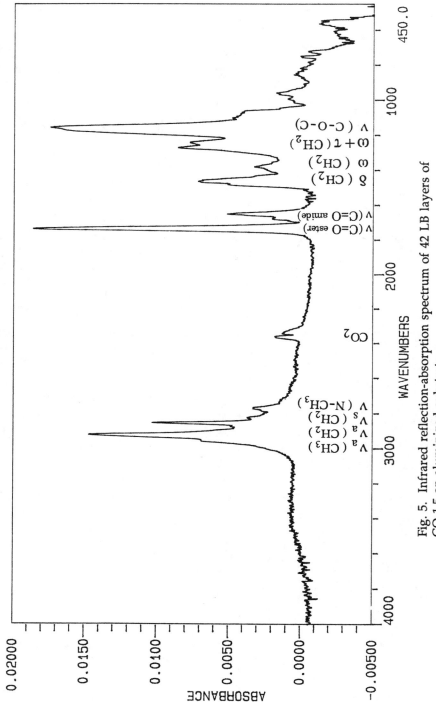

Fig. 5. Infrared reflection-absorption spectrum of 42 LB layers of CO-1.5 on aluminized substrate.

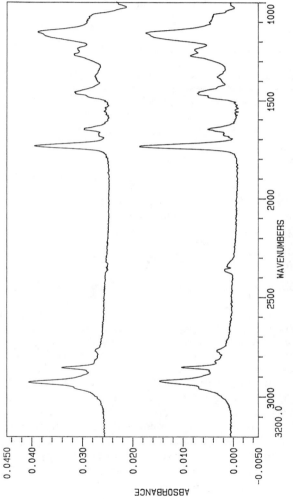

Fig. 6. Temperature dependence of infrared reflection-absorption spectra of 42 LB layers of CO-1.5 on aluminized substrate. Bottom spectrum taken at room temperature (27 °C), top spectrum taken at 46 °C.

temperature of 46 °C. On raising the temperature from 27 to 46°, the spectra show a decrease in the C=O and C-O-C stretch band intensities relative to those of the CH_2 and CH_3 band intensities. Since the bands arising from the alkyl chain (hydrophobic) portion of the chains are unchanged in intensity, we interpret the spectra as indicating that the orientation of the hydrophilic portion of the chain is modified on heating. We are currently investigating this behavior over a wider temperature range to determine the exact orientational change.

Table I: Infrared Band Assignments

Frequency (cm^{-1})	Mode Assignment
2960 w	CH_3 asymmetric stretch
2926 s	CH_2 asymmetric stretch
2853 m	CH_2 symmetric stretch
2770 m	$N-CH_3$ stretch (?)
1736 s	C=O ester stretch
1650 m	C=O amide stretch
1468 m br	CH_2 scissors
1400 m br	$O-CH_2$ wag
1267 m br	$C-CH_2$ wag
1242 m br	CH_2 wag + twist
1157 s br	C-O-C stretch

Mode strength abbreviations: s- strong, m-medium, w-weak, br- broad

Diffusion Measurements. The gas permeability divided by the LB film thickness (P/L) for 18 layers of CO-1.5 on Celgard is shown in Figure 7. (The porous Celgard support does not contribute a significant resistance to gas transport because gas permeates 4 orders of magnitude faster through uncoated Celgard membranes (7)). At room temperature, the P/L for N_2 and CO_2 are nearly identical. At higher temperatures, the P/L ratios remain approximately equal until temperatures above 40 °C are reached. The P/L value for CO_2 increases significantly while the P/L values for N_2 rise less significantly with temperature. The increase of the CO_2 permeability leads to a significant increase in the CO_2 selectivity over N_2 when the temperature is above 40° C as shown in Figure 8.

If an approximate thickness of 50 Å is assigned per LB layer, the CO_2 and N_2 permeability P at room temperature is 4.5 x 10^{-11} cc (STP) / (cm-s-cm Hg). (For a LB film of 18 layers of CO-1.5 the total

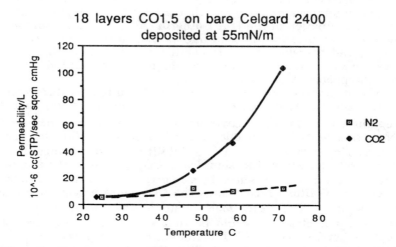

Fig. 7. Temperature dependence of the permeability of asymmetric membrane to N_2 and CO_2. Membrane consists of 20 LB layers of CO-1.5 on Celgard 2400.

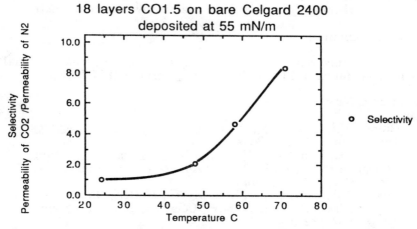

Fig. 8. Temperature dependence of the selectivity of CO_2 over N_2 for asymmetric membrane of 18 LB layers of CO-1.5 on Celgard 2400.

thickness of the LB film is 900 Å). In comparison, the room temperature permeability of CO_2 through amorphous polyethylene is 3.3 x 10^{-9} cc (STP) / (cm-s-cm Hg) (12). The low permeability in the LB film at room temperature suggests that the LB film is highly ordered. From an examination of the Π-A isotherm (13) , the Langmuir layer at the air/water interface is solid-like at the transfer pressure of 55 mN m^{-1} , and therefore the permeability values reflect that of gas permeation in a crystalline film. Presumably the crystallinity is due to the hydrophobic parts of the LB film.

The structural changes in the LB film with temperature should not affect the N_2 permeability as the N_2 solubility is low in both hydrophilic and hydrophobic environments. The nominal increase in the N_2 permeability is solely due to an increase in the diffusion speed (diffusivity) over any decrease in solubility due to an increase in temperature. However, the significant CO_2 permeability increase with temperature may be caused by a reorientation in the hydrophilic parts of the LB film which makes the CO_2 more accessible to the tertiary amine. The exact mechanism of the increase in CO_2 transport is not clear. It could be that structural changes in the hydrophilic portions of the LB films have caused an increase in CO_2 solubility. While the reactivity of CO_2 with tertiary amines is greatly reduced from that of primary or secondary amines, some reactivity remains(14) . Therefore another possibility is that the CO_2 is reacting with the amine groups and that transport of CO_2 in the hydrophilic portion of the LB structure is caused by a "bucket brigade" type mechanism where CO_2 can hop from one reactive amine to another.

Acknowledgments

Funding for this research was provided in part by the National Science Foundation (CBT Division), through grant CBT-8720282.

Literature Cited

1. Gaines, G.L. Insoluble Monolayers at Liquid-Gas Interfaces; 1966, John Wiley and Sons, New York.
2. Stroeve, P.; Franses, E. Molecular Engineering of Ultrathin Polymeric Films; 1987, Elsevier Applied Science Publishers, Crown House, Essex, England.
3. Kowel, S.T.; Selfridge, R.; Eldering, C.; Matloff, N.; Stroeve, P.; Higgins, B.G.; Srinivasan, M.P.; Coleman, L.B. Thin Solid Films, 1987, 152, 377.
4. Roberts, G.G., Adv. Phys. 1985, 34, 475.
5. Albrecht, O.; Laschewsky, A.; Ringsdorf, H.; Journal of Membrane Science, 1985, 22, 187.
6. Rabe, J.P.; Rabolt, J.F.; Brown, C.A. Thin Solid Films, 1985, 133, 153.

7. Stroeve, P.; Coelho, M.A.N.; Dong, S.; Lam, P., Coleman, L.B.; Fiske, T.G.; Ringsdorf, H.; Schneider, J. Thin Solid Films, 1989, 180, 241.
8. Coleman, L.B.; Fiske, T.G.; Stroeve, P.; Coelho, M.A.N.; Dong, S.; Ringsdorf, H.; Schneider, J. Thin Solid Films, 1989, 178, 227.
9. Swalen, J.D.; Rabolt, J.F. "Characterization of Orientation and Lateral Order in Thin Films by Fourier Transform Infrared Spectroscopy", in Fourier Transform Infrared Spectroscopy, 1985, 4, 283.
10. Laschewsky, A.; Ringsdorf, H.; Schmidt, G.; Schneider, J.; Journal of American Chemical Society, 1987, 109, 788.
11. Schneider, J.; Ringsdorf, H.; Rabolt, J.F.; Macromolecules, 1989, 22, 205.
12. Michaels, A.S. and Bixler, H.J.; "Membrane Permeation: Theory and Practice", in Progress in Separation Science, Ed. E.S. Perry , 1968, John Wiley and Sons,. New York.
13. Bruinsma, P., Ph.D. thesis, University of California at Davis, in progress.
14. Astarita, G.; Savage, D.W.; Bisio, A.; Gas Treating with Chemical Solvents,1983, John Wiley and Sons, New York.

RECEIVED August 27, 1990

Chapter 11

Monolayer Structure at Gas–Liquid and Gas–Solid Interfaces

Infrared Reflectance Spectroscopy as a Probe

Richard A. Dluhy[1] and Donald G. Cornell[2]

[1]Department of Chemistry, School of Chemical Sciences, University of Georgia, Athens, GA 30602
[2]Agricultural Research Service, U.S. Department of Agriculture, 600 E. Mermaid Lane, Philadelphia, PA 19118

Within recent years, external reflection-absorbance infrared spectroscopy has been adapted to the study of conformation-sensitive vibrational modes in amphiphilic monolayers at non-metallic surfaces. Included in this group is the class of monomolecular films that form ordered two-dimensional arrays at the air-water (A/W) interface. It has proven possible to investigate the surface-induced first order thermodynamic phase transition of phospholipid monolayer films at the A/W interface using this technique. Results of this work have focused on describing the nature of the phase transition, the design of an automated Langmuir-type film balance and its interfacing to the spectrometer, and the construction of a model to describe the orientation of the vibrational dipole moments based on the polarized reflectance spectra. In addition, the ability to probe the A/W interface allows us to directly compare the IR spectra of *in-situ* monolayers with the spectra of monolayer films transferred to attenuated total reflectance (ATR) crystals using classical Langmuir-Blodgett techniques, in order to determine whether the physical process of transfer disrupts the structure of the monolayer film.

During the decade of the 1980's, the discipline of surface and interfacial science has encountered tremendous growth. Organized two-dimensional organic films at interfaces have become an important component in a wide variety of disciplines, including non-linear optics, sensor applications, catalysis, surface modification, electrode coatings, and biomacromolecules (1). Surface chemistry has also benefited from the development of modern, sensitive spectroscopic techniques that have enabled investigators to study monomolecular films at interfaces spectroscopically for the first time. Table I lists many of the spectroscopic methods that are currently used to study surface and interfaces; several of these techniques were only first developed during the 1980's.

0097–6156/91/0447–0192$06.00/0

Table I. Spectroscopic Techniques Used For Surface Characterization

Ion Spectroscopy

- Secondary Ion MS
- Laser Microprobe MS
- Particle-Induced X-Ray Emission
- Ion Scattering
- Rutherford Backscattering

Electron Spectroscopy

- Low Energy Electron Diffraction
- X-Ray Photoelectron
- Auger Electron
- Electron Energy Loss

Microscopic Techniques

- Electron
- Scanning Tunneling
- Atomic Force

X-Ray Techniques

- Glancing Angle X-Ray Diffraction
- Appearance Potential
- Extended X-Ray Fine Structure

Desorption Techniques

- Electron Stimulated Desorption
- Laser-Induced Desorption
- Thermal Desorption

Optical Spectroscopy

- Fluorescence
- UV-VIS
- Circular Dichroism
- Quasi-Elastic Light Scattering
- Brillouin Scattering
- Second Harmonic Generation
- Sum Frequency Generation
- Raman
- Infrared

Most of the contemporary research areas that utilize surface chemistry techniques employ thin organic films that have been physically or chemically adsorbed onto a solid (usually metallic) substrate. The use of conducting metal surfaces is due not only to their relevance to many different fields, but also to the fact that many surface spectroscopic techniques (Table I) need such a surface in order to produce high-quality spectra. This criterion effectively eliminated many interesting non-metallic surfaces from study using modern surface-sensitive spectroscopic methods.

However, continuing development of spectroscopic sampling techniques in recent years has increasingly led to the use of non-metallic surfaces as thin-film substrates, and the structure of the adsorbed film on these surfaces is now being elucidated via spectroscopic techniques. Included in this group is the class of monomolecular films that form ordered two-dimensional arrays at the air-water (A/W) interface (a full list of abbreviations used in this chapter appears at the end before the literature citations). Insoluble monolayers at the A/W interface have been extensively studied as models for ionic, dipolar, and interfacial phenomena in systems such as surfactants, polymers, proteins, steroids, membranes, and other biophysical systems. The usefulness of these monolayers stems from the fact that the experimenter is able to dynamically change the physical state of the film during the experiment; in addition interfacial interactions can be probed in real time.

Unfortunately, while insoluble monolayers at the A/W interface have been extensively studies as models for interfacial phenomena, there is virtually no information available concerning the detailed physical structure of these films. Historically, this lack of information concerning the molecular-level structure of monolayers at the A/W interface can be traced to the inability of most spectroscopic techniques to study a low surface area, flat water interface with sufficient sensitivity to produce spectra with reasonable signal-to-noise ratios. This situation is changing with the introduction of the modern spectroscopic methods mentioned above. Even here, however, the number of spectroscopic techniques with *in-situ* sensitivity at the A/W interface is not large. Table II lists the methods that have been recently utilized to study the A/W interface *in-situ*.

Table II. Spectroscopic Techniques Used For In-Situ Characterization at the A/W Interface

X-Ray Techniques

• Glancing Angle X-Ray Diffraction

Optical Spectroscopy

• Fluorescence
• Quasi-Elastic Light Scattering
• Second Harmonic Generation
• Sum Frequency Generation
• Resonance Raman
• Infrared Reflectance

From a practical standpoint, the use of the glancing angle X-ray method, while powerful, requires a synchrotron source and therefore, due to the constraints of beam time, is necessarily limited in the number of systems that can be studied in a given time period. Of the optical methods listed, the fluorescence and resonance Raman techniques directly measure spectra of an embedded chromophore, and thus give only indirect information concerning the host monolayer. The second harmonic methods (i.e. second harmonic and sum frequency generation) have the advantage of being uniquely surface-sensitive, however, they are limited to visible detection in the case of second harmonic generation, and suffer from insufficient spectral sensitivity. In contrast, the infrared reflectance method, as we have shown (2-6), is able to give direct information on the structure of the monolayer without the use of probe molecules. In addition, the infrared reflectance technique is highly surface-sensitive, due to the fact that vibrational spectroscopy measures changes in the dipole moment of specific chemical groups in the molecule.

Experimental Methods

Materials. 1,2-Dipalmitoyl-*sn*-glycero-3-phosphocholine (DPPC) was obtained from Avanti Polar Lipids, Inc. (Birmingham, AL) at 99% stated purity. The DPPC was further purified by dissolving it in dry, glass-distilled $CHCl_3$ and reprecipitating the lipid from dry, glass-distilled acetone (5:95 $CHCl_3$:acetone v/v). The precipitate was allowed to cure for one hour under refrigeration before centrifugation. This procedure was repeated six times.

Monolayer Film Preparation. Phospholipid monolayers for the *in-situ* infrared reflectance experiments were formed on the surface of an all-Teflon automated Langmuir film balance, the detailed design of which has been previously reported (3). The demountable trough was prepared for each experiment by cleaning with chromic acid cleaning solution, followed by exhaustive water rinsing. Sub-phase water was prepared from a Milli-Q ion exchange system (Millipore Inc., Bedford, MA) and had a nominal resistivity of 18 mega-ohm/cm. Temperature control of the subphase was established by thermostatting the hollow aluminum base plate on which the trough was mounted. The subphase temperature for the experiments reported here was 22 ± 1 °C. DPPC monolayer films were prepared by spreading from a 1 mg ml^{-1} lipid solution in n-hexane/ethanol (9:1 v/v) onto the A/W interface; the monolayer film was then allowed to equilibrate for ten minutes before the start of data collection. Surface pressure was monitored by use of a filter paper Wilhelmy plate (GVWP, 0.22 μm, Millipore Corp.), which was suspended from an electronic microbalance (Model 27, Cahn Inc., Cerritos, CA). The calculation of surface pressure and molecular area for the pure DPPC monolayer film using these experimental conditions has been described earlier (3).

Infrared Spectroscopy of Monolayer Films. External reflectance IR spectra of the DPPC monolayer film at the A/W interface were obtained using a Digilab FTS-40 FT-IR spectrometer (Digilab, Inc., Cambridge, MA). The IR spectra were collected using 1024 scans at 8 cm^{-1} resolution with triangular apodization, one level of zero-filling, and a narrow-band, liquid-N_2-cooled HgCdTe detector. The incoming radiation was polarized using an IR polarizer (Al wire grid on KRS-5, from Cambridge Physical Sciences); the polarizer was placed in the optical path between the spectrometer and the water surface. For both parallel and perpendicular polarizations, external reflection-absorbance spectra were obtained by ratioing (in absorbance mode) the single beam reflectance spectrum of the monolayer-covered H_2O surface to the single beam reflectance spectrum of pure H_2O which served as the background. Monolayer IR spectra were obtained as a function of surface pressure by compression of the Langmuir film balance barriers until the desired surface pressure was obtained; compression was then stopped during data collection. A barrier compression rate of 0.20 cm min^{-1} was used. Frequencies and intensities of the CH$_2$ stretching modes were calculated using a center-of-gravity algorithm (7); the frequencies of the stretching bands calculated using this method are estimated to be accurate to greater than ± 0.1 cm^{-1}.

Transferred Monolayers. A separate film balance and trough was used for the preparation of Langmuir-Blodgett (L-B) monolayers transferred onto solid substrates. The design of this film balance has been described in detail elsewhere (8). The L-B monolayers were transferred onto germanium ATR crystals (50x10x2 mm, 45° face angle, obtained from Wilmad Glass Company, Buena, NJ). The film balance used for external reflection IR analysis and the film balance used for transferred films both employed Wilhelmy plates and dual compression barriers driving toward the center of the trough. In the preparation of transferred films, the Ge crystal was located centrally between the compression barriers; hence, there was no net movement of the film either at the point where transfer was effected or where the external reflection IR data was obtained. Monolayer films to be transferred were prepared in a manner similar to those used for external reflection IR analysis. After the phospholipid was spread, the solvent was allowed to evaporate for 15 minutes. Compression was begun and proceeded until the desired pressure was reached; transfer was then immediately started. The process of film transfer onto the 50 mm Ge crystal was accomplished with the aid of a rack and pinion device and took approximately 2 minutes. Constant film pressure was maintained during transfer by a microprocessor-controlled feedback loop (8).

Results and Discussion

Monolayer Films at the A/W Interface. Previous studies of phospholipid monolayers at gas-liquid interfaces have shown that it is possible to follow the first order thermodynamic phase transition of these monolayer films using the infrared reflectance techniques described in this manuscript (see e.g. ref. 6 and references cited therein). For long chain hydrocarbon molecules, it has been demonstrated that the frequencies of the antisymmetric and symmetric CH$_2$ stretching vibrations are conformation-sensitive, and may be empirically correlated with the order (i.e. the trans-gauche character) of the hydrocarbon chains (9-11).

   Figure 1 demonstrates such a correlation between the surface pressure of a DPPC monolayer film (Figure 1A) and the frequency of the antisymmetric CH$_2$ stretching vibration (Figure 1B), both measured as a function of molecular area. Four phases are generally described in the pressure-area curve: the liquid-expanded (LE) region (> ca. 76 Å^2 molecule^{-1}), the main liquid-expanded to liquid-condensed (LE-LC) transition region (~55-76 Å^2 molecule^{-1}), the liquid-condensed to solid-condensed (LC-SC) region (~40-55 Å^2 molecule^{-1}) and the collapsed (C) film (< ca. 40 Å^2 molecule^{-1}). The observed band frequency decreases (i.e. the hydrocarbon chains become more ordered) as the molecular area decreases. Previous studies

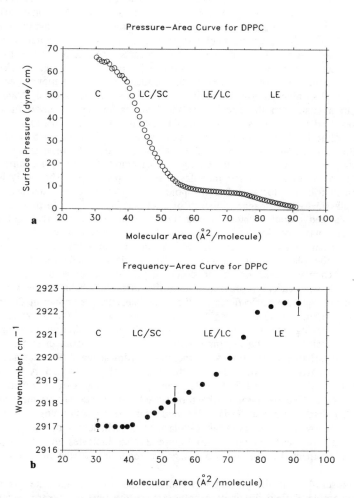

Figure 1.   The surface pressure-molecular area isotherm of the DPPC monolayer at 22 °C (Figure 1A, top) with the infrared frequencies of the CH_2 antisymmetric stretching vibration, plotted against molecular area for the DPPC monolayer (Figure 1B, bottom).

have described this frequency decrease in terms of the thermodynamic state of the monolayer film (6). The large decrease (~ 3 cm^{-1}) in frequency between 55 Å^2 and 75 Å^2 is related to the first-order thermodynamic phase transition of the monolayer. In addition, a second, smaller decrease (~ 1 cm^{-1}) in frequency is noted between 40 Å^2 and 50 Å^2. This frequency shift is not related to the previous phase transition, but rather occurs in the condensed region of the pressure-area diagram where the monolayer film surface pressure begins to rise sharply. This frequency shift can be attributed to a further ordering of the lipid hydrocarbon chains beyond that observed during the main transition. Recent synchrotron X-ray diffraction results also report the appearance of more solid domains in the condensed region of monolayer films of 1,2-dimyristoyl-sn-glycero-3-phosphatidic acid (12). This second transition is assigned to a change in the packing of the hydrocarbon chains upon further film compression and is analogous to a solid-solid type hydrocarbon transition.

In addition to describing the conformation of the hydrocarbon chains for amphiphilic molecules at the A/W interface, external reflectance infrared spectroscopy is also capable of describing the orientation of the acyl chains in these monolayers as a function of the monolayer surface pressure. The analysis of the orientation distribution for an infrared dipole moment at the A/W interface proceeds based on classical electromagnetic theory of stratified layers (2). In particular, when parallel polarized radiation interacts with the A/W interface, the resultant standing electric field has contributions from both the z component of the p-polarized radiation normal to the interface, as well as the x component of the p-polarized radiation in the plane of the interface. The E field distribution for these two components changes based on the incoming angle of incidence of the p-polarized radiation. A geometrical representation of the incoming radiation as well as the resulting mean square electric fields for the A/W interface are shown in Figure 2.

The theoretical absorbances for a monolayer at the A/W interface may be calculated by taking into account this change of E field distribution as a function of the angle of incidence of the incoming radiation. Figure 3 is an example of such a theoretical absorbance calculation for a monolayer film at the A/W interface. In the calculation of Figure 3 it was assumed that the monolayer was 25 Å in thickness, with optical constants of $n=1.5$, $k=0.1$. The optical constants of H$_2$O were taken at 3000 cm^{-1} (13). The behavior of the p-polarized absorbance shows the previously-reported (2) discontinuity at $\sim 54°$ (the value for the Brewster angle of the A/W interface). Below 54°, the expected p-polarized absorbance is negative; above 54° the expected absorbance is positive. Figure 3 also indicates that for s-polarized radiation, the expected absorbances are negative regardless of the incoming angle of incidence.

Figures 4A and 4B present experimental spectra that illustrate the principle that the incoming E field distribution helps govern the type of spectra obtained. In Figure 4A, spectra of a DPPC monolayer are presented which were obtained at 60 ° angle of incidence with s-polarized radiation. As in previous studies where the experimental angle of incidence was 30° (2-6), the observed spectra have negative absorbances. In Figure 4B, however, the spectra of the monolayer taken with p-polarized radiation show positive absorbance bands, as predicted from theory (Figure 3).

In the case of external reflectance at air-metal surfaces, the resultant standing E field has a contribution solely due to the z component of the p-polarized light; the

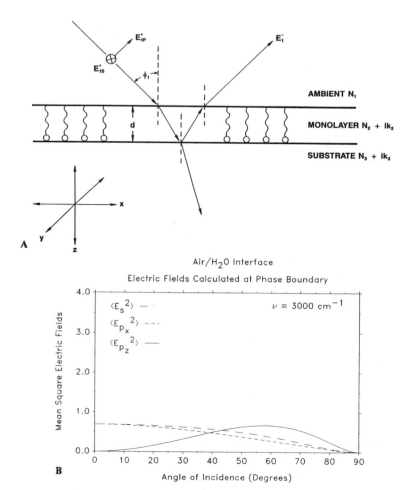

Figure 2.   (A) Schematic illustration of the interaction of a plane electromagnetic wave in a three-phase system.   Parallel (p) and perpendicular (s) components are shown for the incoming propagating wave (+ superscript).   The coordinate geometry is indicated in the lower left of the diagram.   (B) The mean-square electric field intensities at the A/W interface vs. the angle of incidence of the incoming plane wave.   The optical parameters are: $n_1 = 1.0$; $n_2 = 1.441$, $k_2 = 0.0297$, at 3000 cm^{-1}.   The magnitude shown is relative to that of the incident beam (defined as 1.0).   (Figure 2A reproduced with permission from Reference 2.   Copyright 1986 American Chemical Society.   Figure 2B adapted from Reference 2).

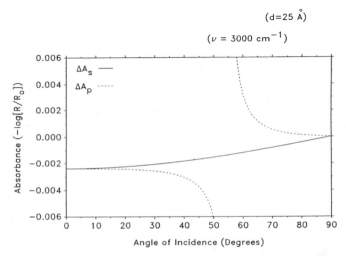

Figure 3. Theoretical absorbance for a monolayer at the A/W interface plotted as a function of the experimental angle of incidence of the incoming radiation. The solid line indicates the theoretical absorbance for s-polarized radiation; the dashed line indicates the theoretical absorbance for p-polarized radiation.

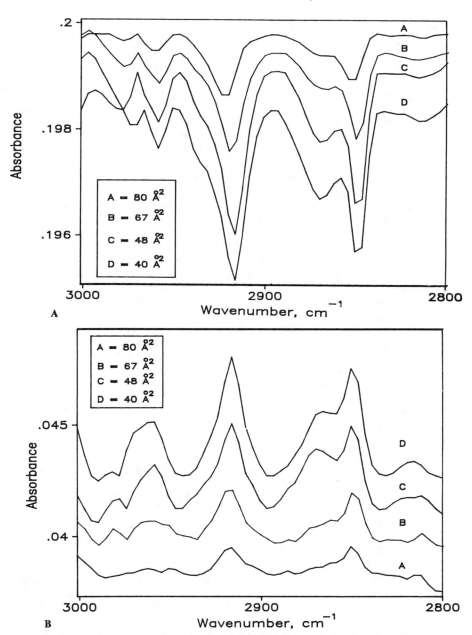

Figure 4.   Experimental spectra of the DPPC monolayer at differing molecular areas.   The angle of incidence of the incoming radiation was 60° relative to the surface normal.   Figure 4A (top) shows s-polarized spectra;   Figure 4B (bottom) shows p-polarized spectra.   The precise monolayer molecular areas at which each spectrum was obtained are given in the figures.

x and y components are negligible due to E field nodes at the interface. For the case of external reflectance at the A/W interface, however, finite values of the E field and, hence, finite absorbances are present in all three orientations (2). This complicates the calculation of an overall dipole orientation distribution since all geometric orientations must be considered.

One approach to a solution of this problem was put forward by Hansen (14), who derived general equations which express the overall experimental film absorbance in terms of the external reflectance of the substrate. These relations contain within them expressions for the individual anisotropic extinction coefficients in each geometric orientation. Solution of these general equations for the anisotropic extinction coefficients allows for an unambiguous description of the dipole orientation distribution when combined with a defined orientation model.

Recently, this type of analysis has been applied to the experimental spectra of DPPC monolayer films at the A/W interface. A detailed derivation and analysis of these equations as specifically applied to the A/W interface will be presented elsewhere (Dluhy, manuscript in preparation). However, Figure 5 presents the results of the calculations for the average orientation angle of the DPPC hydrocarbon chains as a function of molecular area. The orientation angle was calculated at various molecular areas encompassing the LE, LC, and SC regions of the monolayer pressure-area curve. The results presented in Figure 5 show that throughout the LE and LE-LC regions, the average orientation of the DPPC hydrocarbon chains remains relatively constant, at approximately 50-55° from the surface normal. This value may have two interpretations. First, that the hydrocarbon chains of the monolayer, on average, are tilted by 50-55° from the z geometric direction relative to the A/W interface. Alternatively, because of the standard deviation of the measurements, the measured orientations encompass the so-called "magic angle", 54.74°. At this angle in the assumed orientation distribution function all preferred orientations are averaged, with the result being a completely random, isotropic orientation distribution. The closeness of the calculated orientation angles for the hydrocarbon chains of DPPC in the LE and LE-LC regions to this angle precludes us from specifically choosing one alternative.

As the monolayer is further compressed in the condensed phase beyond 50 Å^2 molecule^{-1}, the lipid becomes more oriented until it reaches a point where the hydrocarbon chains of the DPPC molecule are tilted at an average angle of 35° relative to the interface normal. This value agrees well with the value of 30° calculated for the average orientational tilt of the hydrocarbon chains of DPPC monolayers in the condensed phase at the A/W interface using synchrotron X-ray diffraction (15). In addition, this same value (30-35°) has been calculated for the chain tilt of fully hydrated DPPC multilayers in the gel phase (16). Thus, the experimental evidence points to a tightly packed crystal structure for the DPPC monolayer in the condensed phase that is the two-dimensional analog of the three-dimensional low temperature, ordered gel phase structure of the bulk DPPC molecule. However, the IR data suggest that in the expanded and transition phases of the monolayer, DPPC may be in a relatively disordered orientation. This randomness may last until after the LE-LC phase transition, at which point the surface pressure of the condensed monolayer film begins to sharply rise and the hydrocarbon chain becomes more oriented.

Monolayer Films Transferred to Solid Substrates. Historically, the spectroscopic investigation of monolayer physical structure has been performed on films transferred to solid substrates, usually through conventional L-B techniques. A wide variety of methods may then be employed in the study of these films. For example, ultraviolet, circular dichroism, and IR spectroscopy, as well as electron microscopy have been performed on monolayers transferred to quartz (for UV and CD), Ge (for IR), and mica (for EM). While much useful information has been obtained in the study of transferred monolayers, there is always a concern about whether the actual physical process of transfer from a gas-liquid to a gas-solid interface induces a change in the structure of the molecule.

With the development of the external reflectance IR technique for observing monolayers *in-situ* at the A/W interface, we now have the ability, for the first time, to directly compare the structure of the monolayer film at the A/W interface with the monolayer transferred to a solid substrate. In order to determine whether these transfer artifacts occur for the DPPC monolayer, we have studied the structure of DPPC when transferred to Ge ATR crystals. Figure 6 is the pressure-area curve of the DPPC monolayer on which are indicated the points at which film transfer was made. Specific surface pressures of transfer were chosen in order to insure that transferred monolayers were studied in the LE, LE-LC and LC-SC regions, to provide a basis of comparison with the *in-situ* monolayers.

In Figure 7 a comparison is made of the frequency of the CH_2 antisymmetric stretching vibration as a function of molecular area for DPPC monolayer films at the A/W and A/Ge interfaces. As described above, the frequency of this vibration is related to the overall macromolecular conformation of the lipid hydrocarbon chains. For the condensed phase monolayer (~ 40-45 Å^2 molecule^{-1}), the measured frequency of the transferred monolayer film is virtually the same as that of the *in-situ* monolayer at the same molecular area, indicating a highly ordered acyl chain, predominately all-trans in character. For LE films as well as films transferred in the LE-LC phase transition region, however, the measured frequency appears independent (within experimental uncertainty) of the surface pressure, or molecular area, at which the film was transferred. The hydrocarbon chains of these films are more disordered than those of the condensed phase transferred films. However, no such easy comparison can be made to the *in-situ* monolayers at comparable molecular areas. For the LE monolayers ($>$ ca. 70 Å^2 molecule^{-1}), the transferred monolayers are more ordered than the *in-situ* film. In the LE-LC phase transition region (~ 55-70 Å^2 molecule^{-1}), the opposite behavior occurs.

The ATR spectra of the transferred monolayer may also be used to calculate the orientation distribution of the hydrocarbon chains in the transferred film. In this case, the dichroic analysis of the polarized ATR spectra proceeds from well-known principles; a detailed analysis has been presented elsewhere (*17*).

The results for the calculation of the orientation distribution for the hydrocarbon chains in the transferred monolayer films are presented in Figure 8. As is the case with the orientation distribution of the *in-situ* monolayers, the transferred films have a similar tilt angle in the expanded and phase transition regions. For the transferred monolayer, however, the tilt angle is in the range 35-40° from the surface normal, a much more oriented monolayer than the calculations indicate for the *in-situ* film (Figure 5). Figure 8 also shows that the condensed phase transferred monolayers are more oriented than those films transferred in the LE and LE-LC

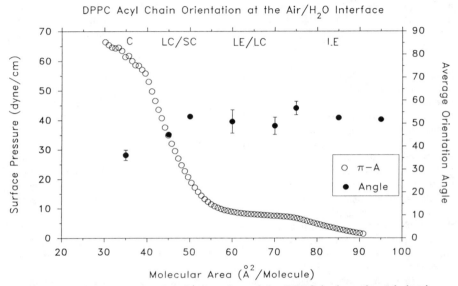

Figure 5. The average angle of orientation of the DPPC hydrocarbon chains in the in-situ film at the A/W interface as a function of the monolayer molecular area (solid circles). The pressure-area curve of DPPC (open circles) is superimposed on the figure.

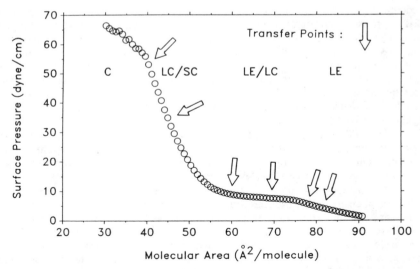

Figure 6. The points along the pressure-area curve of DPPC at which monolayer transfers onto Ge ATR crystals were made.

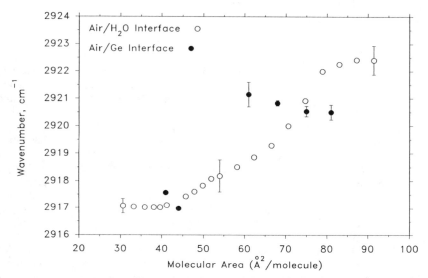

Figure 7. The calculated frequency of the CH_2 antisymmetric stretching vibration in the transferred DPPC monolayer films (solid circles) plotted against the molecular area at which the film was transferred. The frequency of this vibration for the in-situ monolayer film at the A/W interface is superimposed on the plot (open circles).

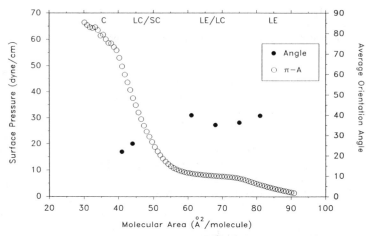

Figure 8. The average angle of orientation of the DPPC hydrocarbon chains in the transferred monolayer films on Ge ATR crystals as a function of the monolayer molecular area (solid circles). The pressure-area curve of DPPC (open circles) is superimposed on the figure.

regions, as well as being more oriented than the *in-situ* monolayer films at similar molecular areas.

Conclusion

Modern methods of vibrational analysis have shown themselves to be unexpectedly powerful tools to study two-dimensional monomolecular films at gas/liquid interfaces. In particular, current work with external reflection-absorbance infrared spectroscopy has been able to derive detailed conformational and orientational information concerning the nature of the monolayer film. The LE-LC first order phase transition as seen by IR involves a conformational gauche-trans isomerization of the hydrocarbon chains; a second transition in the acyl chains is seen at low molecular areas that may be related to a solid-solid type hydrocarbon phase change. Orientations and tilt angles of the hydrocarbon chains are able to be calculated from the polarized external reflectance spectra. These calculations find that the lipid acyl chains are relatively unoriented (or possibly randomly oriented) at low-to-intermediate surface pressures, while the orientation at high surface pressures is similar to that of the solid (gel phase) bulk lipid.

It has also proven possible to directly compare the structure of the monolayer film at the A/W interface with the structure of the monolayer film transferred onto a solid substrate using conventional L-B methods. For DPPC monolayer films transferred to Ge ATR crystals at low-to-intermediate pressures, the transferred monolayer films have a constant conformational order independent of the transfer pressure, and an orientational distribution that is more ordered than that of the *in-situ* monolayer. For those monolayer films transferred at high surface pressures, the hydrocarbon chains have a similar conformational order but are more oriented than the *in-situ* monolayer at the same surface pressure,

Given the ubiquitous nature of interfacial thin films, and the current revival of interest in Langmuir film balance technology, we believe that the results and new techniques described in this chapter will find applications in analytical, physical, and biological chemistry.

Abbreviations Used

A/W	air-water interface
ATR	attenuated total reflectance
C	collapsed
CD	circular dichroism
DPPC	1,2-dipalmitoyl-*sn*-glycero-3-phosphocholine
EM	electron microscopy
FT-IR	Fourier transform infrared
Ge	germanium
IR	infrared
L-B	Langmuir-Blodgett
LC	liquid-condensed
LE	liquid-expanded
SC	solid-condensed
UV	ultraviolet

Acknowledgment

This work was supported by the Public Health Service through National Institutes of Health grant GM40117 (RAD).

Literature Cited

1.   Swalen, J.D., Allara, D.L., Andrade, J.D., Chandross, E.A., Garoff, S., Israelachvili, J., McCarthy, T.J., Murray, R., Pease, R.F., Rabolt, J.F., Wynne, K.J., and Yu, H. Langmuir 1987, 3, 932.
2.   Dluhy, R.A. J. Phys. Chem. 1986, 90, 1373.
3.   Dluhy, R.A., Mitchell, M.L., Pettenski, T., and Beers, J. Appl. Spectrosc. 1988, 42, 1289.
4.   Mitchell, M.L. and Dluhy, R.A. J. Am. Chem. Soc. 1988, 110, 712.
5.   Dluhy, R.A., Reilly, K.E., Hunt, R.D., Mitchell, M.L., Mautone, A.J. and Mendelsohn, R. Biophys. J. 1989, 56, 1173.
6.   Hunt, R.D., Mitchell, M.L., and Dluhy, R.A. J. Mol. Structure 1989, 214, 93.
7.   Cameron, D.G., Kauppinen, J.K., Moffatt, D., and Mantsch, H. Appl. Spectrosc. 1983, 36, 245.
8.   Cornell, D.G. J. Colloid Interface Sci. 1982, 88, 536.
9.   Snyder, R.G., Hsu, S.L., and Krimm, S. Spectrochim. Acta, Part A 1978, 34, 395.
10.   Snyder, R.G., Strauss, H.L., and Elliger, C.A., J. Phys. Chem. 1982, 86, 5145.
11.   MacPhail, R.A., Strauss, H.L., Snyder, R.G., and Elliger, C.A. J. Phys. Chem. 1984, 88, 334.
12.   Kjaer, K., Als-Nielsen, J., Helm, C.A., Laxhuber, L.A., and Mohwald, H. Phys. Rev. Letters 1987, 58, 2224.
13.   Downing, H.D. and Williams, D. J. Geophys. Res. 1975, 80, 1656.
14.   Hansen, W. Symposia of the Faraday Society 1970, 4, 27.
15.   Helm, C.A., Mohwald, H., Kjaer, K., Als-Nielsen, J. Europhys. Letters 1987, 4, 697.
16.   Tardieu, A., Luzzati, V., and Reman, F.C. J. Mol. Biol. 1973, 75, 711.
17.   Cornell, D.G., Dluhy, R.A., Briggs, M.S., McKnight, C.J., and Gierasch, L.M. Biochemistry 1989, 28, 2789.

RECEIVED July 26, 1990

Chapter 12

Adsorption of Proteins and Polysaccharides at Aqueous–Solid Interfaces by Infrared Internal Reflection Spectrometry

In Situ Investigation

K. P. Ishida and P. R. Griffiths

Department of Chemistry, University of Idaho, Moscow, ID 83843

The adsorption of albumin from aqueous solution onto copper and nickel films and the adsorption of ß-lactoglobulin, gum arabic, and alginic acid onto germanium were studied. Thin metallic films (3-4 nm) were deposited onto germanium internal reflection elements by physical vapor deposition. Transmission electron microscopy studies indicated that the deposits were full density. Substrate temperature strongly influenced the surface structure of the metal deposits. Protein and/or polysaccharide were adsorbed onto the solid substrates from flowing (0.5 ml/min) solutions. The net charge on albumin affected how much protein initially accumulated at the aqueous/solid interface; however, essentially the same amount of albumin remained firmly bound to the surface independent of the protein's net charge and the substrate involved. Albumin did not affect the integrity of the copper and nickel films. Protein demonstrated a higher affinity for the solid substrates and was more firmly bound than the polysaccharide material.

When a clean metal surface is exposed to an aqueous environment, bacteria and other microorganisms attach to the liquid/solid interface (1,2). Most adherent bacteria produce extracellular polysaccharides (glycocalyx) that serve to anchor the cells to the surface and to each other (3,4). As the bacteria proliferate, a glycocalyx-enclosed biofilm forms. These biofilms vary in thickness ranging from less than 10 μm to visible gelatinous deposits. Formation of a biofilm on metal surfaces creates a serious fouling problem. Biofilms accumulate on water pipes inside heat exchangers and reduce the efficiency of heat transfer. Biofilm growth also narrows the inner diameter of pipes and reduces their carrying capacity. Biofilms on ships create drag, reduce speed, and increase fuel consumption.

0097–6156/91/0447–0208$06.00/0
© 1991 American Chemical Society

The acidic polysaccharides typically associated with biofilms can exhibit corrosive behavior toward metal surfaces (5-7). Bacteria within the viscous matrix grow, divide, and produce metabolites. The metal surface on which the bacteria grow is ultimately affected by these processes, of which material deterioration and corrosion are possible consequences.

The sequence of events involved in the initial fouling process has been investigated (8), and a two-phase attachment process with a reversible and irreversible attachment phase was described. During the initial reversible phase, bacteria are held to the surface by weak attractions. Firmer binding occurs when physical and chemical forces combine to hold the bacterial cells irreversibly to the surface (e.g. via extracellular polymeric material). The initial fouling process is probably preceded by adsorption of dissolved organic material from the bulk fluid onto the surface (2,9). Only after an organic conditioning film has been acquired are bacteria believed to begin colonizing a surface. These conditioning films have been determined to be primarily of humic- or protein-dominated origin.

Attenuated total refectance (ATR) infrared spectroscopy has been used to identify chemical species that are adsorbed to a germanium internal reflection element (IRE). We have described a method for depositing thin (3-4 nm) films of metal onto cylindrical IREs. The metallic films are sufficiently thin that the evanescent wave penetrates into the bulk aqueous layer making ATR *in situ* studies feasible (10). External reflection has also been used to identify species that adsorbed to metal surfaces, but with a noted loss in sensitivity and spectral contrast compared to ATR (2). The external reflectance data was obtained from a dehydrated metal surface and thus may have contained species that collapsed onto the surface.

In an effort to gain a better understanding of the initial fouling step leading to bacterial attachment and biofilm formation, we have investigated the adsorption of protein and polysaccharides onto thin metallic films and uncoated internal reflection elements from flowing solutions using attenuated total reflectance spectroscopy. The preliminary results will be described in this paper.

Experimental

Bovine serum albumin (BSA), fraction V, ß-lactoglobulin, gum arabic, and alginic acid, low viscosity, were purchased from Sigma Chemical Co. (St. Louis, MO). The proteins were used without further purification. Aqueous solutions of gum arabic and alginic acid were filtered successively through cellulosic membrane filters of 5.0 μm and 0.45 μm pore size to remove insoluble particulate matter. The filtrate was frozen and lyophilized prior to use. Protein content was determined colorimetrically utilizing a BSA standard (11). Protein and alginic acid solutions were prepared in 0.15 N NaCl (saline). Gum arabic solutions were prepared in Milli-Q water. The pH was adjusted with HCl or NaOH.

Copper (99.999%) wire of 0.1 mm diameter and nickel (99.99%) wire of 0.127 mm diameter were purchased from Aldrich (Milwaukee, WI). Thin metallic films (3-4 nm) were deposited by evaporation onto a 0.635-cm diameter x 8.26-cm long germanium cylindrical internal reflection (CIR) element (Spectra-Tech, Stamford, CT) as previously described (*10*). Copper or nickel wire was wrapped around the apex of a V-shaped polished tungsten filament which was mounted 11.1 cm from the IRE. When a vacuum of 10^{-6} Torr was achieved, sufficient current was passed through the filament to melt the metal into a single droplet. The current was gradually increased until the metal disappeared. To achieve a homogeneous coating, the IRE was mounted on a rotary platform and spun at 120 rpm. The effect of substrate temperature on surface morphology and structure was investigated. Metal films were deposited onto heavy carbon-coated Formvar-coated copper grids (Ted Pella, Inc., Redding, CA). Surface continuity and morphololgy of the metal deposits were determined by transmission electron microscopy (TEM). Copper grids were placed on an aluminum heating block inside the vacuum chamber. Copper was deposited onto grids maintained at 146 °C and 370 °C to achieve Zone 2 and Zone 3 structural morphology respectively (*12*), *vide infra*. Nickel was deposited onto grids maintained at 356 °C to achieve Zone 2 structural morphology. Depositions of nickel at a temperature necessary to achieve Zone 3 structural morphology were not accessible by the system and therefore were not studied. Depositions of both metals were carried out at ambient temperature.

The copper- or nickel-coated Ge IRE was mounted in a Pyrex body flow cell (Spectra-Tech) that had a 2-mL cell volume. The flow cell was initially filled with saline or Milli-Q water. Protein and/or polysaccharide solutions, maintained in a 24 °C water bath, were pumped through the flow cell at a rate of 0.5 mL/min and were not recirculated.

Spectra were measured at 4 cm^{-1} resolution with a Nicolet 740 Fourier transform infrared (FT-IR) spectrometer equipped with a medium range mercury-cadmium-telluride detector. A series of 128 scan spectra (43 sec measurement time) was collected every 5 min for the first hour and then every 10 min for 3 hr. At the end of the 4 hr period, saline or Milli-Q water of the same pH was substituted for the polymer solution and the data acquistion program was restarted.

A scaled saline or water reference spectrum was manually subtracted from each sample spectrum such that the region of the spectrum between 1700 cm^{-1} and 2000 cm^{-1} was flat. Each subtracted sample spectrum was baseline corrected between 900 cm^{-1} and 2000 cm^{-1}, and the peak height of characteristic protein and polysaccharide bands was determined using the baseline correct function and the cursor, respectively, on the Nicolet 660 data station.

Characterization of Metal Deposits

The coating process has been viewed as occurring in three steps. First, atoms striking the surface transfer kinetic energy to the substrate and become loosely bound "adatoms". Next, the adatoms diffuse over the surface exchanging energy with the substrate and other adsorbed species until they are either desorbed or become trapped at low energy sites forming stable nuclei. Finally, atoms readjust their positions within the lattice by bulk diffusion processes (*13,14*).

The conformation of a vapor deposited film is influenced by several factors: (1) the nature of the substrate; (2) the temperature of the substrate during the deposition; (3) the rate of deposition; (4) the deposition thickness; (5) the angle of incidence of the vapor stream; and (6) the pressure and nature of the ambient gas phase (*14*).

The microstructure and morphology of thick single-phase films have been extensively studied for a wide variety of metals, alloys, and refractory compounds. Structural models have been proposed (*12,13*). Three zones with different microstructure and surface morphology were described for thick (tens of micrometers) deposits of pure metal. At low temperature (< 0.3 T_m), where T_m is the melting point (K) of the deposit metal, the surface mobility of the adatoms is reduced, and the deposit was reported to grow as tapered crystallites. The surface is not full density (Zone 1). At higher substrate temperature (0.3-0.45 T_m), the surface mobility increases. The surface structural morphology is full density and columnar in nature (Zone 2). Finally, at still higher temperatures (>0.45 T_m) the deposit show equiaxed grain morphology (Zone 3). The effect of temperature on the deposits was investigated due to concern over the effects of radiative heating on the continuity of the deposition, since previous studies on the deposition of copper on germanium IRE indicated the presence of channels in the copper film which exposed the underlying support (*10*). Photomicrographs measured by TEM illustrating the effect of substrate temperature on the nickel and copper deposits on heavy carbon-coated Formvar-coated copper grids are shown in Figures 1 and 2, respectively. The nickel deposit carried out at ambient temperature displayed the finest grain morphology. The surface grain morphology was homogeneous and full density with grain diameters up to 6 nm, as shown in Figure 1a. The grain size increased approximately 5-fold when nickel was deposited onto a heated (T=356 °C) grid, as shown in Figure 1b. Despite the increase in grain size, the underlying Formvar surface was not exposed. Copper grains (averging 15 nm) visible in deposits carried out at ambient temperature were larger than the grains in the nickel deposit carried out under the same conditions. The overall surface morphology of the copper film, shown in Figure 2a, looked similar to the nickel film deposited on grids at 356 °C. The surface morphology of the deposit changed when the grid was heated prior to deposition of copper. At a substrate temperature of 146 °C, individual grains were less noticeable, and the surface appeared more diffuse or opaque. Distinct channels formed throughout the surface, as shown in Figure 2b. The channels were approximately 10 nm in width and appeared to expose the underlying Formvar support. Depositing copper onto a grid heated to 370 °C resulted in an even greater change in surface morphology of the deposit. Large grains (25-30 nm) were visible over the entire surface. Large gaps (on the order of 20-30 nm) exist between these grains, as shown in Figure 2c. The stability of the copper deposits was tested by reheating the copper-coated grids in an annealing process. Grids, coated at ambient temperature and at 146 °C, were heated at 385 °C for 1 hr at a pressure of 10^{-6} Torr. No change in surface morphology was observed for either deposit.

Surface migration during high temperature depositions were apparent from the transmission electron micrographs. The surface morphology changed significantly when nickel was deposited on a substrate heated to achieve Zone 2 type surface morphology. The grain size increased indicating that adatom surface migration occurred prior to condensation. Substrate deposition

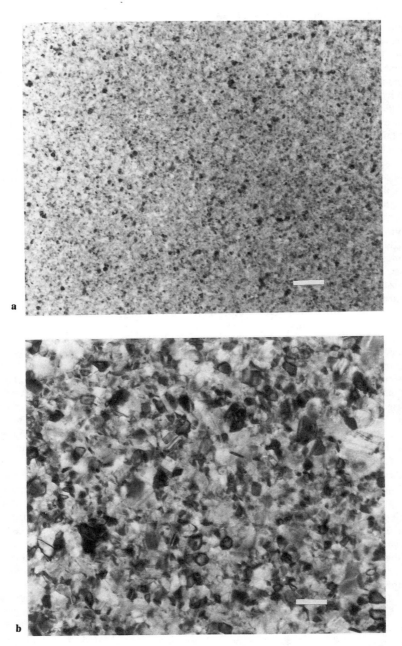

Figure 1. TEM photomicrographs of nickel deposits on a heavy carbon-coated Formvar-coated copper grid coated at (a) ambient temperature and (b) 356 ℃. Key: bar equals 50 nm.

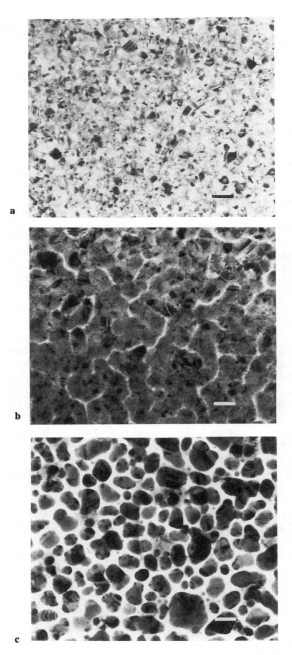

Figure 2. TEM photomicrographs of copper deposits on a heavy carbon-coated   Formvar-coated   copper grid coated at (a) ambient temperature and (b) 146 °C, and (c) 370 °C. Key: bar equals 50 nm.

temperatures greater than 504 °C are necessary to achieve Zone 3 conditions for nickel. Zone 3 conditions were not investigated due to limitations of tne depostion system. Copper deposits were also strongly influenced by the substrate deposition temperature. At a substrate temperature of 146 °C, full density coverage was not achieved. Channels exposing the underlying Formvar support were visible, indicating that extensive adatom migration had occurred at the surface. When the substrate temperature was increased to 370 °C, the effect of adatom migration was most apparent. Distinct nucleated islands formed with large gaps between these islands.

The surface morphology of metal deposits in our study did not resemble those described in the literature; however, our deposits were 3 to 4 orders of magnitude thinner. Three distinctly different surface morphologies were observed with copper indicating that substrate temperature is an important consideration when depositing thin metallic films. Our earlier studies (10) demonstrated the presence of channels in the surface of the copper deposit. This effect probably resulted from surface heating since the tungsten filament was placed 3.8 cm from the surface of the micro cylindrical IRE used in that study, as opposed to the 11.1 cm distance used in our current work. Also, atoms striking the IRE surface under these conditions have greater kinetic energy. Thus, greater surface mobility would be expected. Metals deposited under the present experimental conditions are homogeneous and full density.

Albumin Adsorption on Copper and Nickel Thin Films

0.01% (w:v) solutions of BSA at pHs of 7.4, 4.8 (the isoelectric pH), and 4.0 were flowed past uncoated and coated internal reflection elements. The Amide II band intensity at 1547 cm^{-1} is plotted as a function of time of flow for adsorption of albumin onto germanium and thin metallic films of copper or nickel. The results are summarized in Figure 3. The adsorption isotherms of albumin approximated Langmuir behavior, as the Amide II band rose rapidly and then reached a plateau. The maximum protein adsorption occurred at the isoelectric pH, regardless of the substrate involved. Copper demonstrated a greater ability to accumulate albumin than germanium or nickel at pH 4.8. The intensity of the 1547 cm^{-1} band reached a maximum of 25 mAU for copper, a maximum of 20 mAU for nickel, and a maximum of 18 mAU for germanium. At pH 4.0 copper accumulated more protein than nickel, and nickel slightly more than germanium. The Amide II band plateaued at 16 mAU, 14 mAU, and 13 mAU for Cu, Ni, and Ge respectively. The least amount of protein adsorption occurred at pH 7.4. Significantly more protein adsorbed on copper and nickel than on germanium at pH 7.4, however. Slightly more protein was retained on nickel than on copper. Adsorption isotherms of albumin on nickel at pH 7.4 and pH 4.0 were similar. Under both conditions the Amide II band plateaued at 14 mAU.

The rate of protein adsorption was greatest at pH 4.0, and the rates of adsorption at pH 7.4 and 4.8 were similar. The nature of the substrate did not appear to affect the rate of protein adsorption.

Once adsorbed on a substrate, albumin was firmly bound, as indicated by the minimal change in the intensity of the Amide II band after the saline rinse was initiated. At pH 4.0 and 7.4, very little protein desorbed from the

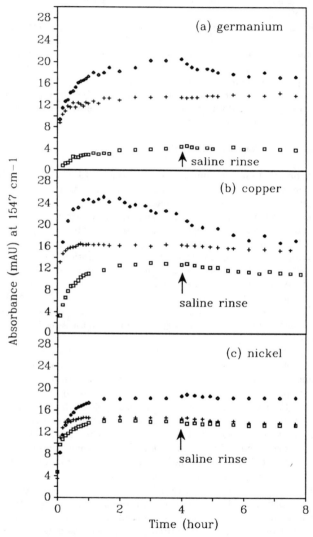

Figure 3. Kinetic plots for 0.01% (w:v) BSA adsorption and desorption as a function of time of flow. Key: (a) germanium, (b) copper, (c) nickel; □ pH 7.4, ◊ pH 4.8, + pH 4.0. (All data not shown for clarity.)

substrate during the 4-hr rinse period. A significant amount of protein desorbed from the copper surface at pH 4.8 before the saline rinse was started. Albumin continued to desorb from the surface during the rinse period; however, the same amount of protein remained firmly bound regardless of the substrate involved.

Albumin did not demonstrate any corrosive behavior to the copper and nickel films. The intensity of the 2100 cm^{-1} water association band, shown in Figure 4, did not change during the 4-hr period that albumin was initially exposed to the thin metal films. Even after 4 additional hours of exposure to the adsorbed protein film, the 2100 cm^{-1} band intensity remained unchanged, indicating that the copper and nickel films remained intact. Exposure of copper and nickel films to albumin above and below the protein isoelectric point did not alter these results.

Gum Arabic Adsorption on Germanium

The filtered preparation of gum arabic contained 13 $\mu g/mg$ protein (dry wt). When a 0.01% (w:v) aqueous solution, at pH 6.5, was exposed to a Ge IRE, no polysaccharide was observed to adsorb from a flowing solution at the aqueous/solid interface, as the characteristic C-O stretching bands of gum arabic were not visible in the water-subtracted spectra. The protein (1.3%) associated with gum arabic demonstrated a high affinity for the Ge surface, adsorbing from a flowing solution at a concentration of 1.3 ppm, while polysaccharide, present at a concentration of 100 ppm, did not adsorb on the IRE. Distinct Amide I and Amide II bands of this adsorbed protein are visible in the water subtracted spectra. At the end of 4 hr, the 1549 cm^{-1} band intensity was 0.9 mAU. Rinsing with Milli-Q water (pH 6.5) did not affect the Amide II band intensity, indicating that the protein was firmly adsorbed to the Ge surface.

At a gum arabic concentration of 0.1% (w:v), protein and a small amount of polysaccharide were detected at the Ge surface as shown in Figure 5. The Amide II band intensity increased slowly and steadily throughout the initial 4-hr period, as shown in Figure 6. The Amide II band intensity peaked at approximately 11 mAU; however, protein adsorption did not stop before the water rinse was initiated. Polysaccharide adsorbed rapidly onto the IRE surface but to a lesser extent than protein. The 1070 cm^{-1} band due to polysaccharide plateaued after 45 min at 1.7 mAU when exposed to the flowing polymer solution. The Amide II band intensity only dropped 10% during the rinse period, indicating that the protein adsorbed on the surface was firmly bound, whereas very little polysaccharide remained adsorbed on the IRE surface. The 1070 cm^{-1} band intensity dropped rapidly and then stabilized when the biopolymer-coated IRE was exposed to the water rinse. Half of the polysaccharide material adsorbed initially remained associated with the IRE surface at the end of the rinse period. Neverless, these experiments indicated that at equilibrium about three times as much albumin adsorbed to the IRE than alginic acid. Estimates of the absorptivity of the analytical bands from KBr disk spectra of alginic acid and albumin showed that the absorptivity of the Amide II band of albumin is about twice as great as that of the 1034 cm^{-1} band of alginic acid.

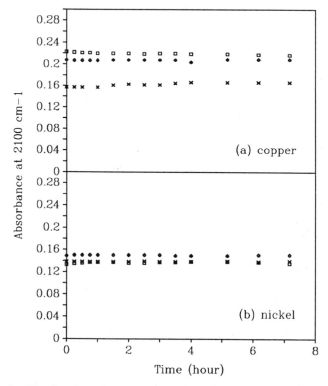

Figure 4. Kinetic plots demonstrating the effect of 0.01% (w:v) BSA on thin metal films based on 2100 cm^{-1} band intensity. Key: (a) copper, (b) nickel; □ pH 7.4, x pH 4.8, ◊ pH 4.0. (All data not shown for clarity.)

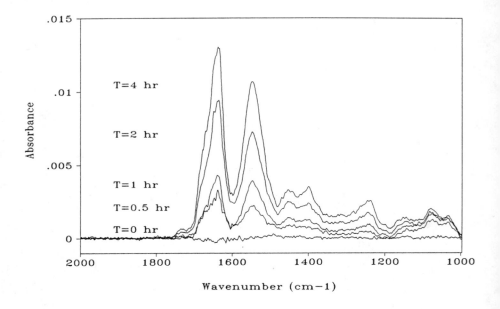

Figure 5. ATR spectra of material adsorbed from flowing 0.1% (w:v) gum arabic, pH 6.5, onto a germanium ATR crystal after 0, 0.5, 1.0, 2.0, and 4.0 hours after the start of the experiment.

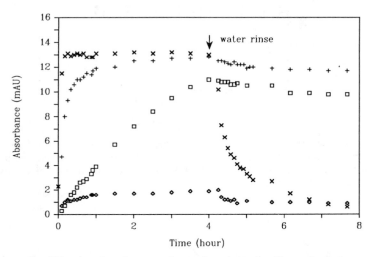

Figure 6. Kinetic plot for protein and polysaccharide adsorption and desorption as a function of time of flow of gum arabic solutions at pH 6.5. Key: 0.1% (w:v) gum arabic, □ 1549 cm^{-1}, ◊ 1035 cm^{-1},; 1.0% (w:v) gum arabic, + 1549 cm^{-1}, x 1035 cm^{-1}. (All data not shown for clarity.)

At a gum arabic concentration of 1.0% (w:v), polysaccharide was detected at the Ge surface within the first minute of exposure to the flowing solution as shown in the first spectrum (T=0 min) in Figure 7. The intensities of both the protein and polysaccharide bands then rose rapidly in a 15-min period of time as shown in Figure 6. The 1070 cm^{-1} polysaccharide band plateaued at 14.7 mAU after 30 min whereas the Amide II band stabilized at 12.5 mAU after approximately 2 hr. The intensity of the 1070 cm^{-1} polysaccharide band dropped rapidly when Milli-Q water was pumped through the flow cell. A 90% decrease in the 1070 cm^{-1} band intensity occurred over the 4-hr rinse period. The final intensity of this band was not significantly different from the intensity observed when gum arabic was adsorbed onto germanium at a concentration that was 10 times less. The protein was more firmly bound to the IRE surface as indicated by the Amide II band intensity which dropped less than 10% during the rinse period. Only 15% less protein remained firmly attached to the Ge IRE when it was adsorbed from a gum arabic solution concentration of 0.1% as compared to 1%. Experiments to study adsorption of proteins and polysaccharides on copper and nickel are not yet complete, but appear to show similar trends.

Alginic Acid Adsorption on Germanium

The protein contamination (1.9 μg/mg, dry wt) in alginic acid was significantly less than that in gum arabic, making alginic acid more suited for model studies. Alginic acid adsorbed rapidly onto the Ge surface from a 1% solution at pH 7.4 as evidenced by the change in the 1034 cm^{-1} band intensity with time of flow; the C-O stretching band plateaued at approximately 18 mAU in 20 min, see Figure 8. Desorption of the polysaccharide occurred at a slower rate and to a lesser extent when compared with gum arabic at the same concentration (1.0%). The 1034 cm^{-1} band intensity dropped 60% during the 4-hr rinse period, as shown in Figure 8. No evidence of the coadsorption of protein was observed in this experiment.

Alginic Acid Adsorption in the Presence of a Protein Conditioning Film

To investigate the effect of preadsorbing a protein film on the adsorption of polysaccharide, ß-lactoglobulin was allowed to adsorb onto germanium from a dilute solution (0.01% w:v). The Amide II band intensity plateaued at approximately 10.5 mAU, as shown in Figure 9. No protein desorbed from the surface during the 4-hr rinse period. An alginic acid solution (1.0% w:v) was then exposed to the adsorbed protein film, and the polysaccharide adsorbed rapidly onto the protein/germanium matrix. The 1034 cm^{-1} band plateaued in a shorter period of time as compared with adsorption of alginic acid onto a bare Ge IRE. Slightly more polysaccharide adsorbed onto the "protein conditioned" Ge surface when compared with the bare IRE surface. The adsorbed protein layer was not affected by the adsorption of alginic acid from solution, as indicated by the absence of any change in the Amide II band intensity over the 4-hr period.

The protein/polysaccharide system was allowed to remain stagnant for approximately 52 hr (T=16 hr to T=67.5 hr) and was then rinsed for 4 hr

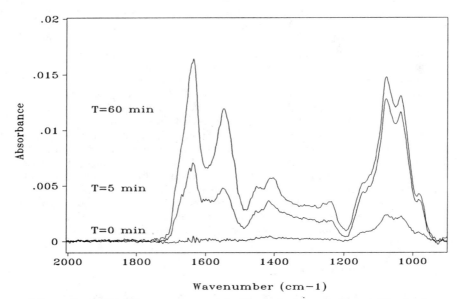

Figure 7. ATR spectra of material adsorbed from flowing 1.0% (w:v) gum arabic, pH 6.5, onto a germanium ATR crystal after 0, 5 and 60 minutes after the start of the experiment.

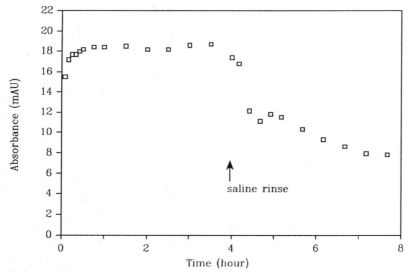

Figure 8. Kinetic plot for 1.0% (w:v) alginic acid, pH 7.4, adsorption and desorption as a function of time of flow. Key: □ 1034 cm $^{-1}$. (All data not shown for clarity.)

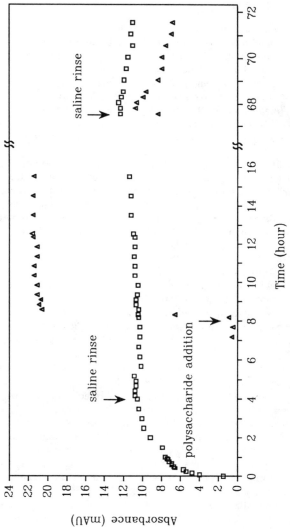

Figure 9. Kinetic plot for 0.01% (w:v) ß-lactoglobulin and 1.0% (w:v) alginic acid adsorption on and desorption from germanium ATR crystal as a function of time of flow. Key: □ 1548 cm^{-1} band of ß-lactoglobulin; △ 1034 cm^{-1} band of alginic acid. (All data not shown for clarity.)

with saline. Alginic acid rapidly desorbed from the protein/polysaccharide matrix, while ß–lactoglobulin remained firmly adsorbed to the germanium surface. The 1034 cm^{-1} band intensity dropped to 8 mAU after the saline rinse. The same quantity of polysaccharide remained firmly bound to the protein "conditioned" and bare germanium surfaces.

Discussion

The metal film (3–4 nm) thickness on the germanium IRE is approximately two orders of magnitude less than the penetration depth of the electric field; and therefore, the germanium, metal, protein/polysaccharide multiphase system can be treated like the "thin film" case described by Harrick (16).

At the low protein concentration (0.01% w:v) used in our studies, the signal from protein can be assumed to be entirely due to adsorbed material. Very little or no change in the Amide II band intensities occurred when protein in the bulk phase was replaced by saline or water during the rinse period. Protein remaining after the saline rinse can be considered as firmly adsorbed material.

The net charge on albumin appears to be more significant than the nature of the substrate when considering how much protein initially binds to the aqueous/solid interface. More protein adsorbed onto copper, nickel, and germanium substrates at pH 4.8, where albumin has no net surface charge, than at pH 4.0 or pH 7.4. Since no charge effects exist between the macromolecules adsorbed on the surface, high protein densities at the aqueous/solid interface would be expected. Copper and nickel appeared to accumulate the same quantities of albumin independent of the pH studied.

A rough surface presents more surface area on which physical absorption can occur. More protein might have been expected to adsorb on the copper and nickel films as compared to germanium due to the granular nature of these films. Only at pH 7.4 did copper and nickel accumulate more albumin than did germanium, however. Despite the macroscopic differences in surface morphology copper and nickel appeared to accumulate the same quantity of albumin.

The rate of protein adsorption onto copper and nickel was greatest at pH 4.0. Surface charge effects are probably the main factor determining adsorption rate. The positively charged protein macromolecules are attracted to the solid substrates which are typically negatively charged (8). The amino acid cystine, in concentrations found *in vivo*, has been demonstrated in an electrochemical study by Svare *et al.* to strongly influence copper and nickel passivation (15). In the case of copper, passivation was enhanced, while in the case of nickel the presence of cystine strongly inhibited passivation. Alanine and bovine serum albumin had little or no effect on the anodic dissolution rate of copper and nickel in the study of Svare *et al.* (15).

In our spectroscopic study, where no potential was applied to the metal, albumin did not appear to exhibit a corrosive effect on the thin metal films. If metallic copper or nickel is ionized or solubilized and removed from the surface, the metallic layer will decrease in thickness. Metals absorb strongly in the mid–infrared; and thus, a decrease in the thickness of this absorbing layer would result in an increase in the penetration depth of the evanescent wave (*16*). All band intensities, including the 2100 cm^{-1} water association band, would be expected to increase in intensity if the metallic films were solubilized. The 2100 cm^{-1} band intensity did not change when exposed to albumin for an extended (8 hr) period of time. Altering the pH of the protein solution also did not affect the results.

The extracellular polysaccharides produced by bacteria are believed to be composed of acidic and neutral sugars. Gum arabic and alginic acid were selected to simulate a bacterial glycocalyx. Gum arabic is composed of arabinose, rhamnose, galactose, and glucuronic acid. Alginic acid is composed entirely of the acidic sugars mannuronic and guluronic acid. Despite efforts to remove contaminating materials, protein still remained in both preparations. The protein in gum arabic, which has been reported to be covalently linked to the polysaccharide (*17,18*), demonstrated a very high affinity for the solid substrates, adsorbing from a bulk solution concentration of 1.3 ppm. The high surface activity of proteins is well known. Adsorption of plasma proteins onto biomaterials has been studied extensively (*19-21*). Proteins associated with cellular interactions with foreign materials are present in plasma at concentrations ranging from 0.1-60 mg/ml (*22*). Yet, even at a solution concentration two orders of magnitude less we have demonstrated that protein still adsorbs tenaciously to solid surfaces. The fact that protein (0.19 μg/ml) in the alginic acid preparation did not adsorb onto Ge may give some indication of the concentration limit above which sorption will occur. Neither gum arabic nor alginic acid was retained at the surface as tenaciously as the protein film. Polysaccharide concentration had no affect on the quantity retained at the aqueous/solid interface. On the other hand, the presence of a protein conditioning film did affect how much polysaccharide was retained at the IRE surface. It should be noted that quite different protein and polysaccharide adsorption phenomena were observed previously in our laboratory (*23*). When gum arabic (10% w:v) was exposed to a copper-coated IRE under stagnant conditions, both protein and polysaccharide were detected within 20 min of the initial exposure of the copper film. After rinsing with distilled water, half of the polysaccharide material remained adsorbed to the copper surface while protein was not retained. Obviously further studies are indicated, and are continuing in our laboratory. Studies in the future will be aimed at determining how the protein conditioning film influences the corrosive effects of the acidic polysaccharides.

Acknowledgments

This work was supported by a grant from the National Science Foundation No. DMR-8900417. We would like to thank John Kitasako, Department of Plant Pathology, University of California, Riverside for the preparation of the electron micrographs.

Literature Cited

1. Marszalek, D. S.; Gerchakov, M. S.; Udey, L. R. *Appl. Environ. Microbiol.* **1979**, *38*, 987.

2. Baier, R. E.; Mayer, A. E.; DePalma, V. A.; King, R. W.; Fornalik, M. S. *J. Heat Transfer* **1983**, *105*, 618.
3. Fletcher, M.; Floodgate, G. D. *J. Gen. Microbiol*. **1973**, *74*, 325.
4. Costerton, J. W.; Geesey, G. G.; Cheng, K.-J. *Sci. Am.* **1977**, *238*, 86.
5. Geesey, G. G.; Mittelman, M. W.; Iwaoka, T.; Griffiths, P. R. *Materials Performance* **1986**, *25*, 37.
6. Nivens, D. E.; Nichols, J. M.; Henson, J. M.; Geesey, G. G.; White, D. C. *Corrosion* **1986**, *42*, 204.
7. Jolley, J. G.; Geesey, G. G.; Hankins, R. B.; Wright, R. B.; Wichlacz, P. L. *Surface and Interface Analysis* **1988**, *11*, 371.
8. Marshall, K. C. *Interfaces in Microbial Ecology*; Harvard University Press: Cambridge, MA, 1976; p 36.
9. Loeb, G. I.; Neihof, R. A. In *Applied Chemistry at Protein Interfaces*; Baier, R. E., Ed.; ACS Symposium Series No. 145; American Chemical Society: Washington, DC, 1975, pp 319-335.
10. Iwaoka, T.; Griffiths, P. R.; Kitasako, J. T.; Geesey, G. G. *Appl. Environ. Microbiol.* **1986**, *40*, 1062.
11. Lowery, O. H.; Rosenbrough, N. J.; Farr, A. L.; Randall, R. J. *J. Biol. Chem.* **1951**, *198*, 265.
12. Movchan, B. A.; Demchishin, A. V. *Fizika Metall* **1969**, *28*, 653.
13. Thorton, J. A. *Ann. Rev. Mater. Sci.* **1977**, *7*, 239.
14. Bunshah, R. F.; Thorton, J. A. *Deposition Technologies for Films and Coatings*; Noyes Publications: Park Ridge, NJ, 1982; 83-243.
15. Svare, C. W.; Belton, G.; Korostoff, E. *J. Biomed. Mater. Res.* **1970**, *4*, 457.
16. Harrick, N. J. *Internal Reflection Spectroscopy*; Harrick Scientific Corporation: Ossining, NY, 1987; 13-65.
17. Anderson, D. M. V.; In *Gums and Stabilizers for the Food Industry*; Phillips, G. O.; Wedlock, D. J.; Williams, P. A., Eds.; IRL Press, Oxford, Vol. 4; pp 31-37.
18. Vandevelde, M. C.; Fenyo, J. C. *Carbohydr. Polym.* **1985**, *5*, 251.
19. Gendreau, R. M.; Leininger, R. I.; Winters, S.; Jakobsen, R. J. In *Biomaterials: Interfacial Phenomena and Applications*; Copper, S. L.; Peppas, N. A.; Hoffman, A. S.; Ratner, B. D., Eds.; ACS Symposium Series No. 199; American Chemical Society: Washington, DC, 1982; pp. 371-394.
20. Gendreau, R. M.; Jakobsen, R. J. *Biomed. Mater. Res.* **1979**, *13*, 893.
21. Wang, S. W.; Lee, R. G. In *Applied Chemistry at Protein Interfaces*; Baier, R. E., Ed.; ACS Symposium Series No. 145; American Chemical Society: Washington, DC, 1975, pp 218-229.
22. Horgett, T. A. In *Biomaterials: Interfacial Phenomena and Applications*; Copper, S. L.; Peppas, N. A.; Hoffman, A. S.; Ratner, B. D., Eds.; ACS Symposium Series No. 199; American Chemical Society: Washington, DC, 1982; pp. 233-244.
23. Geesey, G. G.; Iwaoka, T.; Griffiths, P. R. *J. Colloid. Interface Sci.* **1987**, *120*, 370.

RECEIVED August 2, 1990

Chapter 13

Enzymes Adsorbed onto Model Surfaces

Infrared Analysis

Gloria M. Story, Deborah S. Rauch, Philip F. Brode, III,
and Curtis Marcott

Miami Valley Laboratories, The Procter and Gamble Company, P.O. Box
398707, Cincinnati, OH   45239–8707

Adsorption of the enzymes subtilisin BPN' and lysozyme onto
model hydrophilic and hydrophobic surfaces was examined
using adsorption isotherm experiments, infrared reflection-
absorption   spectroscopy   (IRRAS),   and   attenuated   total
reflectance   (ATR)   infrared   (IR)   spectroscopy.   For both
lysozyme   and   BPN',   most   of   the enzyme   adsorbed onto the
model   surface   within   ten seconds.   Nearly an   order-of-
magnitude   more BPN'   adsorbed on   the hydrophobic Ge surface
than   the hydrophilic one, while lysozyme adsorbed somewhat
more strongly to the hydrophilic Ge surface.   No changes in
secondary   structure   were   noted   for   either enzyme.   The
appearance   of carboxylate   bands   in some of   the adsorbed
BPN'   spectra   suggests   hydrolysis   of   amide   bonds   has
occurred.

Although   X-ray   crystallography,   NMR,   and   circular dichroism   are
extremely   valuable   techniques   for   determining   the   structure   of
crystalline   proteins or proteins in solution, they cannot be used to
study   proteins   adsorbed   on   surfaces.   Vibrational   spectroscopy
(infrared   and Raman) appears to   be the best approach   available for
bridging the gap between adsorbed proteins and proteins in solution.
The problem with infrared studies of adsorbates on solid surfaces
is   that the bulk solid also contributes strongly to the IR spectrum.
To avoid this problem, thin model surfaces were prepared to represent
the   bulk solids without adding large absorbances   to the IR spectra.
The   goal is to design the model system   so the interface of interest
dominates   the   IR   spectrum.   Although   model surfaces   more closely
representing actual surfaces of interest would be more desirable, for
simplicity   we started with more general   hydrophilic and hydrophobic
model surfaces.
Adsorption   isotherms   are   an   effective   means   to   measure the
quantity   of enzyme adsorbed as a function of bulk concentration.   In
addition,   the   shape   of   the   isotherm gives   an indication of the
affinity   of the enzyme for the solid   surface (1,2).   By determining

0097–6156/91/0447–0225$06.00/0
© 1991 American Chemical Society

the projected area per molecule on the surface, we can obtain an estimate of molecular packing at the interface. These packing results are an indirect way to obtain evidence of autolysis or denaturation at the interface. Classical adsorption isotherm measurements do not, however, provide several key pieces of information which are more readily obtained by infrared investigations. When using batch methods of substrate–adsorbate equilibration and separation, it is not possible to observe rapid equilibration kinetics. Also, the evidence for autolysis or denaturation of the enzyme on the surface is indirect and dependent on an assumed molecular orientation at the interface. Infrared studies are capable of monitoring these phenomena directly as they occur at the solid/liquid interface. In addition, some confirmation of the quantitative results of the adsorption isotherm experiments is provided.

With infrared reflection absorption spectroscopy (IRRAS), it is possible to obtain information about the orientation of enzyme molecules adsorbed on flat metal surfaces (3,4). Electric dipole-transition moments oriented perpendicular to a flat metal surface show enhanced IR absorbance. IR bands due to vibrations of groups with transition moments oriented parallel to the surface are not observed. The IR-beam component which is polarized perpendicular to the plane of incidence (parallel to the surface) contains no information and can be eliminated by using a polarizer.

Gold mirror substrates modified with thin coatings can be made either hydrophilic or hydrophobic. If the coating is thin compared to the wavelength of light, the metal surface selection rule will still hold (5). SiO_2-coated gold mirrors were used to model the high–surface–area quartz powder used in the adsorption isotherm studies of subtilisin BPN'. These model surfaces can be cleaned and annealed in order to render them hydrophilic. Reaction of these surfaces with dichlorodimethylsilane will render them hydrophobic. The hydrophilic and hydrophobic mirrors prepared in this way were then soaked in BPN' solutions and studied by IRRAS. Differences in the amount of enzyme adsorption and possible orientation can then be determined.

Attenuated total reflectance (ATR) infrared spectroscopy (6) is another useful technique for studying adsorbed proteins (7,8). The cylindrical internal reflectance (CIRcle) cell is ideally suited for aqueous solution infrared studies. Using the CIRcle cell, we can measure IR spectra of adsorbed enzymes with high signal-to-noise ratio (S/N) in aqueous solutions as dilute as 0.01% total protein. The depth of penetration of the IR beam at each reflection is less than 1 μm over the entire mid–IR range with a germanium (Ge) internal reflection element (IRE) (5,6). The total effective path length of the cell is therefore less than 15 μm for a standard–sized CIRcle cell with ~15 reflections. Under these conditions, the intensity of the water band at 1645 cm^{-1} is less than 1 absorbance unit, making subtraction of the spectrum of liquid water possible. The small depth of penetration of IR radiation into the liquid phase makes the CIRcle cell extremely sensitive to the molecules at the surface of the IRE. In this work, parallel studies of the enzymes subtilisin BPN' (30% alpha helix, 20% beta sheet, 15% turns, and 35% random structure) and chicken egg lysozyme (41% alpha helix, 11% beta sheet, 21% turns, and 27% random structure) will be reviewed.

Experimental

All of the infrared experiments were performed on a Digilab FTS-40 Fourier transform infrared (FT-IR) spectrometer equipped with a narrow-band liquid-nitrogen-cooled mercury-cadmium-telluride (MCT) detector. The spectrometer was operated at a nominal resolution of 4 cm^{-1} using a mirror velocity of 1.28 cm/s. The data collected using the gas chromatography (GC) IR software were measured at 8 cm^{-1} resolution. Protein assays for all the experiments were measured on a Beckman DU-70 UV-visible spectrophotometer.

Adsorption Isotherm Experiment. Crystalline quartz (SiO$_2$) in the form of Berkeley MIN-U-SIL 5 and MIN-U-SIL 30 were supplied by Pennsylvania Glass Sand Corp., whose metal oxide analysis yields 99.7% SiO$_2$. This nonporous powdered solid was acid cleaned with repeated washes of concentrated HCl. Reaction with dichloro-dimethylsilane converted the very hydrophilic surface of clean SiO$_2$ to a hydrophobic methylated surface. Covalent bonds are formed with the SiO$_2$ surface through elimination of HCl. The Si-OH groups on the quartz surface and the -SiCl groups of the dichlorodimethylsilane react to produce Si-O-Si bonds, with the dimethyls of the silane at the newly modified interface. Surface areas were determined on these powdered solids using the BET (Brunauer, Emmett, Teller) N$_2$ adsorption isotherms measured on a Quantasorb Jr. (Quantachrome). The hydrophilic MIN-U-SIL 5 has a specific surface area of 5.06 m^2/g. Acid-cleaned MIN-U-SIL 30 has a surface area of 2.16 m^2/g which decreases to 1.81 m^2/g upon methylation.

Subtilisin BPN' was prepared through a series of protein purification steps applied to the fermentation broth. These steps included: ultrafiltration; ethanol precipitation; DEAE (diethyl-aminoethyl) Tris Acryl batch anionic exchange; SP (sulfopropyl) Tris Acryl column cationic exchange; and, concentration with an Amicon stirred cell. The enzyme purity was determined to be ~95% via spectroscopic assays that measure the ratio of active enzyme to total protein. In addition, purity was verified via HPLC and SDS-page (sodium dodecyl sulfate polyacrylamide gel electrophoresis).

Adsorption isotherms were measured by the solution depletion technique. Enzyme solutions were equilibrated with powdered quartz for varying lengths of time. The solid was removed by centrifugation and the supernatants were analyzed. Total protein content of the solutions, before and after equilibration with the quartz, was measured via absorbance at 280 nm and the difference calculated as adsorbed protein. All adsorption isotherms were measured at 25.0°C in Tris buffer (0.1M Tris[hydroxymethyl]aminomethane and 10 mM CaCl$_2$ adjusted to pH 8.6 with HCl).

IRRAS Experiment. The IRRAS accessory used in this work was a modification of one half of a Wilks Scientific double-beam ATR attachment designed for use on a dispersive IR spectrometer. All measurements were made in an atmosphere of dry nitrogen. A gold wire-grid polarizer was placed between the MCT detector and the IRRAS accessory to remove the perpendicularly-polarized component of the infrared beam which contains no information about adsorbed monolayer films. The gold-mirror substrates, 1" X 3/4" optically flat glass slides with 150-200 nm of vapor-deposited gold, were obtained from

Brysen Coating Laboratories. Some of the gold-mirror substrates were purchased with a 10-nm layer of SiO_2 on them. The SiO_2-coated mirrors were cleaned and annealed to produce model hydrophilic surfaces. Half of these slides were then reacted with dichloro-dimethylsilane to produce model hydrophobic surfaces. Using a clean gold mirror for a reference, IR reflection-absorption spectra of these model surfaces were collected both before and after each step in the preparation of the model surfaces. Finally, the model surfaces were soaked in BPN' solutions stabilized with Tris buffer. For comparison, a clean gold substrate was soaked in a 0.1% lysozyme in 0.15 N NaCl solution (pH 7.2). After pulling the substrate out of this solution at a slow, constant rate of speed with a motor, spectra were collected both before and after rinsing with deionized water. All spectra were collected after the surfaces had dried.

CIRcle Experiment. The CIRcle cell used for these experiments is a high-pressure standard cell from Spectra Tech with removable Teflon tubing to allow the cell to be filled in the spectrometer sample compartment without disturbing the N_2 purge or the cell alignment. It can be fitted with a thermal jacket and water bath for temperature studies. The Ge IRE used for these experiments was modified to have a hydrophilic surface by placing it in a Harrick plasma-cleaner, supported with glass wool, for ten minutes. A model hydrophobic surface was produced by dipping the IRE in a 0.5% atactic polystyrene in toluene solution for thirty minutes. The IRE was removed from the solution using a motor at a slow, constant rate to ensure an even coating. The IRE was then inserted and sealed to the cell body wall with Teflon o-rings, and the residual polystyrene removed from the conical ends of the IRE with a toluene-soaked swab. Using a Hummer Nugget sputter coater (Anatech, Ltd.) with a 60% Au/40% Pd target, we deposited a Au/Pd coating onto the Teflon o-rings prior to assembling the CIRcle cell to substantially reduce interfering Teflon bands in the spectra. The o-rings were tilted in four different orientations and a 50-nm layer of Au/Pd was deposited at each position. A final layer was deposited on the back side. Once the cell was assembled and aligned in the sample compartment of the spectrometer, openings around the transfer tubing were sealed off with plastic and masking tape to preserve the N_2-purged environment. A 500-scan single-beam spectrum of the empty cell was collected as a reference. The cell was then filled with the same salt solution used for the enzyme. After a 500-scan absorbance spectrum of the blank salt solution was collected, the gas chromatography (GC) IR data collection software was started for a five minute run. Interferograms were collected at 4 scans/second with individual IR spectra files being created every second. At approximately one minute into the run, 2 mL of an enzyme solution was injected through the transfer tubing into the cell. Once the initial 5-minute kinetics run was completed, a sequence of hourly 500-scan absorbance spectra were collected. Sometimes spectra were collected after the cell was rinsed with blank solvent, leaving only the enzyme that adsorbed to the IRE. The cell would then be purged overnight with a stream of nitrogen and a spectrum of the dried residue would be collected.

Proteins readily adsorb to surfaces, including the IRE of the CIRcle cell. After every experiment, the IRE and cell body needed to be thoroughly cleaned. All contaminated parts were scrubbed with a paste of cerium oxide (CeO) and MICRO solution (a mixed-surfactant detergent manufactured by International Products Corp.). All the scrubbed parts, with the exception of the IRE, were placed in a beaker full of MICRO solution and heated for thirty minutes in a 60°C water bath. The IRE was placed in a beaker lined with tissue, covered in MICRO solution, and heated for thirty minutes in a 60°C water bath. The cell body parts were rinsed in deionized water (DIW), then ethanol, acetone, and hexanes and finally dried with a stream of N_2. The IRE was rinsed with DIW, scrubbed with more CeO/MICRO paste, rinsed with DIW, ethanol, acetone, and hexanes, and finally dried and stored in tissue until used.

Results and Discussion

The adsorption isotherms of subtilisin BPN' on hydrophilic SiO_2 and hydrophobic methyl-SiO_2 are shown in Figure 1. Adsorption on these two surfaces is very different, as reflected in both the magnitude and shape of the two isotherms. At every bulk concentration more BPN' is adsorbed on the hydrophobic than on the hydrophilic surface. Using the hydrodynamic radius for this globular protein obtained from light-scattering studies (Sullivan, J. F., The Procter & Gamble Company, personal communication) we can estimate the projected area per molecule on the surface. Comparing the area per molecule for the plateau region of both isotherms provides an indication of the packing of the enzyme on each surface. The plateau of the hydrophobic isotherm corresponds to adsorptive packing of approximately 65% of a monolayer, based on the projected surface area (2000 Å²) required to adsorb an intact molecule. In contrast, the hydrophilic surface only adsorbs the equivalent of about 15% of a monolayer in the plateau region.

The affinity of BPN' for the two surfaces is also in sharp contrast. Even at very low concentrations, the enzyme's strong affinity for the hydrophobic surface causes an abrupt rise in adsorption. The isotherm for the hydrophilic surface shows a very gradual rise in adsorption as the bulk concentration is increased.

The enzyme adsorption-isotherm experiments on high-surface-area quartz powder were compared to IR reflection-absorption results on model SiO_2-coated gold mirrors. The spectra taken at various stages of the surface preparation revealed how unstable the SiO_2-coated mirrors were to atmospheric contamination and the cleaning and annealing procedure. The SiO_2 bands would often shift and even disappear. The successfully cleaned and annealed SiO_2-coated mirrors were reacted with dichlorodimethylsilane to produce a model hydrophobic surface. This treatment results in absorbances at 1263, 1099, 1024, and 810 cm^{-1}. These are the major absorption bands in polydimethylsiloxane and are clear evidence that the surface has been methylated. A hydrophilic and a hydrophobic mirror were soaked in a 1.2 mg/mL BPN' solution containing Tris buffer to stabilize the enzyme. After drying, the spectrum of the hydrophilic surface revealed no enzyme present. The spectrum of the hydrophobic surface revealed a strong absorbance at 1598 cm^{-1} with shoulders at 1645 and 1537 cm^{-1}. This large band is likely due to the Tris buffer used to stabilize the BPN' in solution. The shoulders are probably due to the amide I and II bands of BPN', respectively. These data correlate with the adsorption isotherms which suggest that BPN' adsorbs about seven times more on a hydrophobic surface than a hydrophilic one at this concentration. Due to all the interferences in the spectrum, no orientation information could be reliably determined.

A clean gold mirror substrate that had been dipped in a 0.1% lysozyme solution yielded a spectrum with amide I and II bands at 1672 and 1543 cm^{-1}, respectively. Rinsing the mirror with deionized water and drying removed about two-thirds of the enzyme from the surface. Discounting interferences from the Tris buffer bands, the amide I and II region of the BPN' and lysozyme IRRAS spectra are very different. This could be due to either inherent differences in the two enzymes or be an artifact of drying the enzymes. The fact that the lysozyme experiment was done on a gold surface while the BPN' experiment was performed on SiO$_2$-coated gold may also be significant. IRRAS measurements in the presence of solvent performed in an electrochemical cell (9) could eliminate the possibility that the enzymes denatured during drying. However, we opted to test this using the CIRcle cell solution ATR technique because it is a more straightforward experiment.

In order to test whether our CIRcle cell spectra were dominated by adsorbed protein or protein in solution, we ran spectra of a series of lysozyme solutions ranging in concentration from 0.1% to 10%. The IR response of the amide I and II bands at 1653 and 1543 cm^{-1} is nearly linear with concentration between 5 and 10% lysozyme. However, the IR intensities change very little between 0.1 and 1%, strongly suggesting that most of the signal we observe at 0.1% concentration is due to adsorbed lysozyme. Since our study of subtilisin BPN' was done at 0.01%, we are almost certainly observing only adsorbed species in our ATR spectra.

Figure 2 shows an overlay of the solution ATR, dry ATR, and dry IRRAS spectra of lysozyme. In each case, the original solution was 0.1% lysozyme in 0.15 N NaCl (pH 7.2). The solvent spectrum has been subtracted from the solution ATR spectrum in Figure 2. The dry ATR spectrum was collected after the cell had been rinsed once with solvent and dried with a stream of nitrogen. The apparent shifts in the amide I, II, and III (1300-1200 cm^{-1}) regions, observed upon comparing the wet and dry CIRcle cell spectra, suggest a conformational change has occurred. The large shift in the amide I band (up to 1670 cm^{-1}) for the IRRAS spectrum is believed to be largely due to an optical effect. For strongly absorbing surface species, the refractive index of the surface is decreased somewhat on the high-wavenumber side of the absorption band. This causes the surface to be less reflective at that wavelength, leading to an apparent increase in the wavenumber of the peak maximum (10).

From the adsorption-isotherm experiments, it is shown that BPN' adsorbs about twenty-times more on hydrophobic quartz than on hydrophilic quartz at 100 ppm. It is also known that BPN' autolyzes readily, especially on hydrophobic surfaces (Brode, III, P. F. and Rauch, D. S., The Procter & Gamble Company, unpublished results). For the first BPN' experiments, the Ge IRE was used directly after cleaning. "Clean" Ge has a water contact angle of about 45°, somewhere between hydrophobic and hydrophilic. To avoid spectral interferences from the Tris buffer bands, the buffer was eliminated from the BPN' solutions used for the CIRcle cell experiments. Only CaCl$_2$ (10 mM final concentration) was used to stabilize the BPN' against autolysis, and the solution was adjusted to a pH of 8.6 using

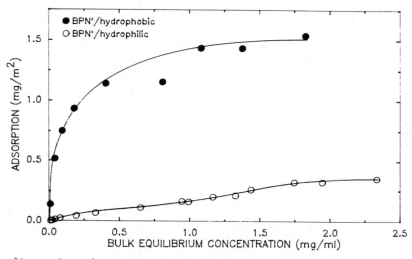

Figure 1. Adsorption of subtilisin BPN' onto hydrophobic and hydrophilic quartz powder at 25.0°C, 10 mM $CaCl_2$, and pH 8.6.

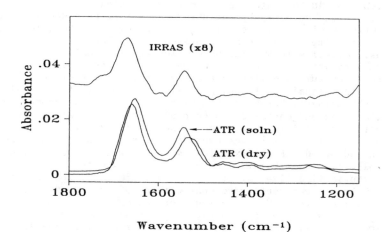

Wavenumber (cm⁻¹)

Figure 2. Overlay of the solution ATR, dry ATR, and dry IRRAS spectra of 0.1% lysozyme. The solvent spectrum has been subtracted from the solution ATR spectrum.

a dilute NaOH solution. The autolysis rate of BPN' is lower in dilute solutions and the enzyme activity, checked before each experiment, ranged from 90 to 100%. After a blank CaCl$_2$ solvent spectrum was collected, a 100 ppm BPN' solution was injected into the cell and spectra were collected approximately every twenty minutes for three hours. The spectra, after blank solvent subtractions, revealed not only the amide I and II bands, but a gradual increase in band intensities at about 1580 and 1400 cm^{-1} (see Figure 3). These bands suggest a slow rise in carboxylate functional components. The carboxylate bands indicate that amide linkages have been hydrolyzed by BPN'. The amide bands remained unchanged in both band shape and position. This suggests that the rapid loss of BPN' activity (determined by assaying the solution in the cell) is due to autolysis and not necessarily conformational changes (denaturation). It is not known, however, whether the adsorbed protein observed initially is intact BPN' or autolyzed fragments.

Through this experiment, it was also noted that coating the o-rings with Au/Pd did not eliminate the Teflon interferences. Apparently, the Au/Pd coating either abrades off during assembly or it doesn't completely prevent the evanescent IR beam from penetrating slightly into the Teflon o-rings. As BPN' is adsorbed around the IRE and o-rings, less of the evanescent wave penetrates into the Teflon and negative Teflon bands "grow" into the spectra as time progresses. Thus, the time and effort required to coat the o-rings appear to be of no benefit. Careful centering of the IRE in the cell body can help to minimize Teflon bands in the spectra.

A set of comparison studies were performed on both lysozyme and BPN', using hydrophilic and hydrophobic model surfaces with both uncoated and Au/Pd-coated Teflon o-rings. Using the GC-IR software, any net changes detected by the MCT detector are recorded graphically in a Gram-Schmidt (GS) "chromatogram" (change in the interferogram signal vs. time). This plot represents the change in the total IR response relative to a reference collected with only the solvent in the cell before injection of the enzyme solution. Since an entire infrared spectrum is collected every second, windows can be set up to monitor individual functional groups. As a result, functional group (FG) "chromatograms" (absorbance within a certain wavenumber window vs. time) can be generated to supplement the information in the GS reconstruction (11–14). A window set up to monitor the amide II region, 1590–1480 cm^{-1} was used to estimate the total amount of protein adsorbed. Eight experiments were performed using the GC-IR data-collection software. Three 500-scan spectra were collected after the initial five minute GC-IR software run, spaced one hour apart. Four experiments were done with a 190 ppm lysozyme in 0.15 N NaCl solution (pH 7.2), and four were done with a 100 ppm BPN' in 10 mM CaCl$_2$ solution (pH 8.6). The conditions for each set of experiments were: 1) a hydrophilic model surface (plasma-cleaned Ge) with and without Au/Pd coating the Teflon o-rings; and 2) a hydrophobic model surface (polystyrene-coated Ge) with and without Au/Pd coating the Teflon o-rings. Figures 4 and 5 show graphs (amide II absorbance vs. time) that summarize all eight experiments. Figure 4 on the lysozyme experiments, indicates that lysozyme adsorbs somewhat more readily to a hydrophilic surface than a hydrophobic surface. Coating the Teflon o-rings appears to have little effect on lysozyme adsorption to a hydrophobic surface and a somewhat

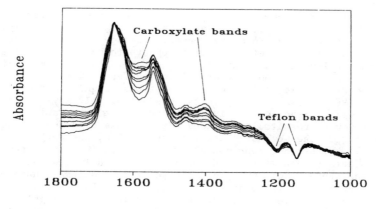

Figure 3.   A series of spectra of 100 ppm subtilisin BPN' from 6 to 180 minutes after injection into the CIRcle cell.   The solvent spectrum has been subtracted.

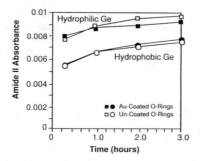

Figure 4.   Adsorption of 190 ppm lysozyme onto hydrophobic and hydrophilic Ge IREs as determined by the amide II band absorbance after subtraction of the solvent spectrum.

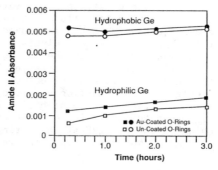

Figure 5.   Adsorption of 100 ppm subtilisin BPN' onto hydrophobic and hydrophilic Ge IREs as determined by the amide II band absorbance after subtraction of the solvent spectrum.

detrimental effect on the adsorption to a hydrophilic surface. The corresponding BPN' experiments are shown in Figure 5. Clearly, BPN' adsorbs more readily to a hydrophobic surface than a hydrophilic surface. Coating the Teflon o-rings appears to have little effect on BPN' adsorption to a hydrophobic surface, but it dramatically enhances adsorption on a hydrophilic surface. When Au/Pd-coated Teflon o-rings were used with BPN' solutions, significant etching of the Ge IRE at the o-ring site was noted. It may be that an electrochemical reaction involving the Au/Pd, Ge, the $CaCl_2$ solution, and the enzyme results in enhanced adsorption around the o-ring site and etching of the Ge IRE. Autolysis products of the enzyme are also detected, but are not reproducible. It is possible that sometimes the products remain at the IRE surface and are detected and at other times the products disperse into the solution and remain undetected. Figures 6 and 7 are overlays of the GS and FG profiles of the hydrophilic lysozyme experiment (uncoated Teflon o-rings) and the hydrophobic BPN' experiment (uncoated Teflon o-rings) respectively. Plots such as these were key to understanding the adsorption characteristics of the enzymes during the first five minutes after injection. The effects of different surface properties on the initial adsorption kinetics of an enzyme can easily be determined by comparing GS or FG profiles. The FG profile, which focuses in on the amide II band, is a more sensitive probe of surface adsorption than the GS profile. It was found for both lysozyme and BPN', that most of the enzyme adsorbed onto the model substrate within ten seconds after injection. Lysozyme continues to adsorb onto the surface at a very slow rate after the initial rapid adsorption, while BPN' maintains its initial coverage.

Conclusion

Using a cylindrical internal reflectance (CIRcle) cell and GC–IR data collection software, it was determined for both lysozyme and BPN', that most of the enzyme adsorption occurred within ten seconds after injection. Nearly an order-of-magnitude more BPN' adsorbed on the hydrophobic surface than the hydrophilic one, while lysozyme adsorbed somewhat more strongly to the hydrophilic Ge surface. Over time periods of about one day, the lysozyme layer continued to increase somewhat in thickness, while BPN' maintained its initial coverage.

From the BPN' adsorption isotherms, the coverage is estimated to be about one-third of a monolayer at this concentration on hydrophobic surfaces. No secondary conformational changes were noted for either enzyme. The appearance of carboxylate bands in some of the adsorbed BPN' spectra suggests some hydrolysis of amide bonds has occurred. The Au/Pd coating applied to the o-rings of the CIRcle cell in an attempt to eliminate the interfering Teflon absorption bands from the spectra had a significant effect on the adsorption of BPN'. An apparent electrochemical reaction occurred, involving BPN', Ge, Au/Pd, and the salt solution used to stabilize BPN'. The result of this reaction was enhanced adsorption of the enzyme around the coated o-rings, etching of the Ge IRE at the o-ring site, and some autolysis of the enzyme. No such reaction was observed with lysozyme.

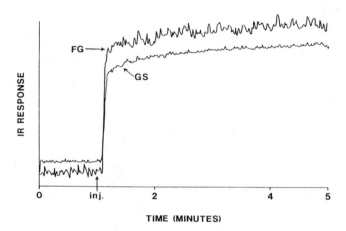

Figure 6. Overlay of the Gram-Schmidt (GS) and functional group (FG) profiles of the adsorption of 190 ppm lysozyme on hydrophilic Ge (uncoated o-rings).

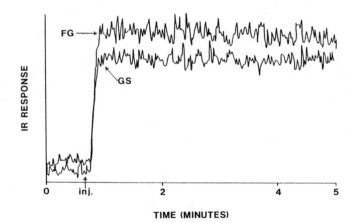

Figure 7. Overlay of the Gram-Schmidt (GS) and functional group (FG) profiles of the adsorption of 100 ppm subtilisin BPN' on hydrophobic Ge (uncoated o-rings).

Acknowledgment

We would like to thank Robert Jakobsen for several helpful discussions.

Literature Cited

1. Norde, W.; Fraaye, J. G. E. M; Lyklema, J. In Proteins at Interfaces. Physicochemical and Biochemical Studies; Brash, J. L.; Horbett, T. A., Eds.; ACS Symposium Series No. 343; American Chemical Society: Washington, D. C., 1987; pp 36-47.
2. Andrade, J. D. In Surface Interfacial Aspects of Biomedical Polymers, Protein Adsorption; Andrade, J. D., Ed.; Plenum Press: New York, NY, 1985, Vol. 2; p 1-80.
3. Greenler, R. G. J. Chem. Phys. 1966, 44, 310.
4. Greenler, R. G. J. Chem. Phys. 1969, 50, 1963.
5. Marcott, C. In Metals Handbook, Ninth Edition; American Society of Metals: Metals Park, OH, 1986, Vol. 10; pp 109-125.
6. Harrick, N. J. Internal Reflection Spectroscopy; John Wiley & Sons: New York, NY, 1967.
7. Gendreau, R. M.; Leininger, R. I.; Winters, S.; Jakobsen, R. J. In Biomaterials: Interfacial Phenomena and Applications; Cooper, S. L.; and Peppas, N. A., Eds.; ACS Advances in Chemistry Series No. 199; American Chemical Society: Washington, DC, 1982; p 371.
8. Gendreau, R. M. In Spectroscopy in the Biomedical Sciences; Gendreau, R. M., Ed.; CRC Press, Inc.: Boca Raton, FL, 1986; pp 21-52.
9. Bewick, A.; Pons, S. In Advances in Infrared and Raman Spectroscopy; Hester; Clark, Eds.; Hayden and Sons: London, 1985, Vol. 12; pp 1-63.
10. Allara, D. L.; Baca, A.; Pryde, C. A. Macromolecules 1978, 11, 1215.
11. Bio-Rad, Digilab Division GC-32 Operator's Manual M091-0329B; Bio-Rad, Digilab Division: Cambridge, MA.
12. deHaseth, J. A.; Isenhour, T. L. Anal. Chem. 1977, 49, 1977.
13. Griffiths, P. R.; deHaseth, J. A.; Azarraga, L. V. Anal. Chem. 1983, 55, 1361A.
14. Chittur, K. K.; Fink, D. J.; Leininger, R. I.; Hutson, T. B. J. Colloid Interface Sci. 1986, 111, 419.

RECEIVED August 22, 1990

Chapter 14

Measurement of Monolayers Adsorbed on Mica
Infrared Techniques

D. A. Guzonas, M. L. Hair, and C. P. Tripp

Xerox Research Centre of Canada, 2660 Speakman Drive, Mississauga, Ontario, L5K 2L1, Canada

Several techniques have been developed to obtain infrared transmission spectra of adsorbed monolayers of organic material on the surface of mica. The bands due to adsorbed monolayers are weak in intensity because of the low surface area of the mica and these bands are obscured by interference fringes arising from the multiple reflections from the back and front surface of the mica. The problem of interference fringes has been removed by using very thin mica sheets or by performing the experiments *in situ*. An improvement in the detection limit has been accomplished by increasing the signal from the monolayer by increasing the amount of sample in the beam using a multiple film technique or by reducing the noise level of the FTIR by operating in a dual beam configuration.

Micas are minerals consisting of a two dimensional sheet of XO_6 octahedra (with X being usually Mg^{2+}, Fe^{2+} or Al^{3+} cations) sandwiched between two sheets of SiO_4 tetrahedral in a 2:1 structure (1). Hydroxyl groups lie at the common plane of the tetrahedral and octahedral layers (Figure 1). The substitution of every fourth Si in the tetrahedral layers by Al gives the surface of the sandwich a charge which is neutralized by bound cations. The weak cation sandwich bonds can be easily broken, giving true micas excellent cleavage along the 100 plane. In muscovite mica, the intersurface cation is potassium, while the octahedral layer consists of AlO_6 octahedra with every third cation site empty. The slight mismatch in the sizes of the faces of the SiO_4 tetrahedral and the XO_6 octahedra leads to a monoclinic crystal structure for mica. As a consequence of this, mica is biaxial (has two optical axes) and has three different refractive indices. Only along one of the optical axes is mica optically isotropic.

The tendency of mica to cleave easily along the (100) crystallographic plane results in smooth, clean reproducible surfaces. This property, coupled with good thermal stability, has led to its wide use as a substrate in surface chemical studies (2). Indeed, the ability of Muscovite mica to be cleaved to give atomically smooth surfaces over areas of several square centimeters

0097–6156/91/0447–0237$06.00/0
© 1991 American Chemical Society

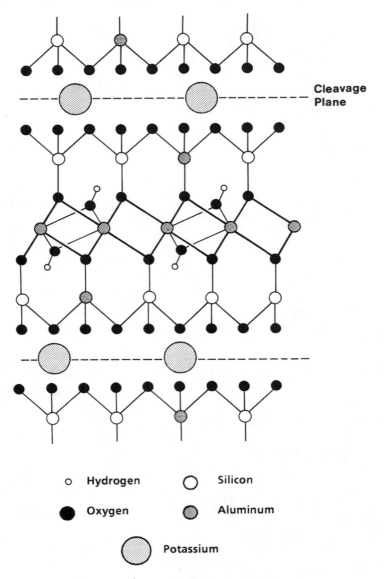

Figure 1. Structure of Muscovite Mica

has led to its near universal use in the surface force apparatus (*3*). In the surface force apparatus, the force exerted between two mica sheets is measured as a function of the separation between the two surfaces. The sensitivity of the measurement (changes in force of 10^{-7} Newtons and of 0.1 nm in distance) has facilitated the use of this device in the study of interfacial forces at the molecular level. We are particularly interested in the adsorption of polymers and surfactants on mica (*4-5*). The detailed comparison of the measured force-distance curves for adsorbed polymers with theoretical predictions requires knowledge of the amount of adsorbed material. It is possible to obtain quantitative measurements of the adsorbed amount from refractive index changes in the surface force apparatus. However, refractive index measurements using the surface forces apparatus are difficult to make for many polymer systems because of the small differences in refractive index between solvent and polymer. Furthermore, interpretation of the refractive index measurements are often complicated by factors such as the presence of two (or more) different blocks in the polymer and the possibility of preferential solvent uptake in mixed solvent systems.

The surface forces technique measures the force between molecules (eg. surfactants, polymers) adsorbed on mica sheets. In the case of large molecules such as polymers, the measurement is most sensitive to the regions closest to the solution and provides little direct information about the region adjacent to the surface. As it is a measurement between macroscopic surfaces, it is unable to provide information on microscopic chemical differences at the interface. Infrared spectroscopy could provide additional information about the quantity of adsorbed material on the mica surface, the identity and orientation of the adsorbed species, and possibly the nature of the surface linkage.

Infrared spectroscopy has been a common tool for the study of solid surfaces (*6*). As in any surface spectroscopy, the number of adsorbed molecules and the surface area of the solid determines the sensitivity needed for IR studies. For low area surfaces, reflection techniques have been used to measure IR spectra of adsorbed monolayers on metal surfaces (*7*). However, for nonmetallic surfaces such as mica, the low reflectivity of mica makes reflection techniques less suitable for IR measurements. At the same time, the biaxial properties of mica, the parallel nature of the surfaces, and the absorbance of the mica itself present difficulties in IR spectroscopy (*8*).

In particular, large interference fringes from the parallel front and back surfaces of the mica sheet obscure the weak absorbance due to the adsorbed monolayers. In FTIR, one measures an optical interference pattern (interferogram) which, for a broad spectral source, is characterized by an intense central region which tapers off rapidly. The reflection of the infrared beam from the front and back surface of the mica produces a secondary interferogram which appears as a spike at a fixed distance from the primary interferogram. The presence of this secondary spike in the interferogram results in a sinusoidal oscillation (fringing) being produced in the spectrum upon performing the Fourier transform. The period of the oscillation in the spectrum is related to the thickness of the mica by the following equation (*9*):

$$D = 1/(2 \, \eta v \cos \theta) \qquad (1)$$

where D is the mica thickness, η is the index of refraction of mica, ν is the period of the interference fringe and θ is the angle between the surface normal and the infrared beam. For sample thicknesses of around 100 μm this periodicity is ~30 cm⁻¹ and the fringes are of about the same width as the bands due to adsorbed species. As they are usually several orders of magnitude more intense it is difficult if not impossible to distinguish between these bands and the fringes.

Several ingenious optical and digital techniques have been used to reduce the fringe problem (7, 9-10). In a recent paper, Carson and Granick (8) were successful in recording infrared spectra of adsorbed octadecyltrichlorosilane (OTS) on mica in transmission. In their study, they used thin mica sheets of about 50 - 200 μm in thickness which gave rise to an oscillation with a period of ~ 50 cm⁻¹. The effect of this fringing was reduced by using parallel polarized light incident at the Brewster's angle (to minimize the reflectivity of the mica) and along one of the optical axes of the birefringent mica. Although successful in reducing the fringing problem, the authors noted that their technique was very sensitive to the relative orientation of the mica with respect to the polarizer and that a small residual oscillation (with an amplitude of about 10⁻³ absorbance) was always present. Since the intensity of the bands due to adsorbed monolayers on mica fall between 10⁻³ to 10⁻⁵ absorbance the residual fringe amplitude may still dominate the spectrum. For comparison to surface force measurements it would also be useful to carry out infrared measurements in the presence of a solvent, a procedure which is difficult to envisage using the arrangement of Carson and Granick. We have recently published an alternative method for the reduction of the fringing problem in the IR spectroscopy of mica, involving the use of thinner mica sheets in transmission (11), and we have used this method and a number of additional experimental methods to circumvent the fringing problem and to improve our sensitivity for monolayer detection.

Experimental

A Bomem Michelson 102 FTIR equipped with a CsI beamsplitter and DTGS detector was used to collect spectra. Spectra were collected at 4 cm⁻¹ resolution requiring approximately 6 seconds per scan and processed using Spectra-Calc software on a PC AT type system.

The dual beam spectrometer used a Bomem Michelson 110–E equipped with a KBr beamsplitter and midrange (650 cm⁻¹) MCT (measured $D^* = 2 \times 10^{10}$) detector (details to be published). Spectra were recorded at 4 cm⁻¹ resolution and each scan required 1 second to collect. Data collection and processing were performed as above.

Cyclohexane, toluene, and CCl_4 were obtained from commercial sources and were distilled prior to use. OTS (Aldrich) was used as obtained. The Polyethyleneoxide/polystyrene (PEO/PS) diblock copolymer ($M_w(PEO) = 4000$, $M_w(PS) = 100,000$, $M_w/M_n = 1.4$) was characterized by GPC and NMR.

The signal-to-noise ratio (S/N) is a very important factor in measuring the usefulness of a particular technique for measuring monolayers on mica. Most of the spectra shown are of very short scan time so that S/N can be clearly measured and compared. Since S/N improves with (scan time)$^{\frac{1}{2}}$ the S/N levels for long scans can be easily predicted from the spectra.

Transmission Studies

The infrared spectrum of mica recorded in transmission is shown in Figure 2. Bands between 1250 and 300 cm^{-1} have been assigned to Si-O-Si, Si-O-Al, Al-O-Al, Si-O, Al-O stretching and AlO-H bending vibrations, 2200 to 1250 cm^{-1} to various overtone and combination modes, while the bands between 3750-3500 cm^{-1} are assigned to Al-OH stretching vibrations (*12-13*). It is apparent from Figure 2 that mica provides a wide transparent optical window between 3500 and 1250 cm^{-1} for transmission IR studies. Since mica can be cleaved to various thicknesses it is possible to circumvent or reduce the effect of the fringing by controlling the thickness of the sheets. By using very thick micas (~ 2 mm) the fringe period becomes smaller than the instrument resolution and disappears. This method is not practical for adsorption measurements because the strong absorption from the mica itself overwhelms the surface contribution. Alternatively, by going to very thin micas (~ 1 μm) the fringe period is much greater than the bandwidth of bands due to adsorbed monolayers.

To exploit this latter alternative, mica sheets were cleaved from larger crystals of Muscovite mica until a thickness of the order of 1-5 μm was reached as judged by the brightness and color of the fringes seen under room light. Actual film thickness was determined from the infrared fringe spacing. Mica at these thicknesses is delicate but can be handled if care is taken: it is possible to cleave mica to thicknesses below 1 μm but at these thicknesses the mica becomes too delicate to use unsupported. The cleaved mica sheets were mounted in standard transmission holders constructed from Teflon.

Several of the advantages of going to a thin mica sheet can be seen in Figure 2. Figure 2b shows the absorbance spectrum of a mica sheet which is about 70 μm thick as measured by the fringe spacing. The inset shows an expansion of the small region around 2900 cm^{-1} showing the magnitude of the observed fringes. The period of this fringing (45 cm^{-1}) is comparable to that reported by Carson and Granick (*8*) and, since a monolayer of OTS would have an intensity of ~ 2 x 10^{-3} adsorbance in the C-H stretching region, it is apparent that the absorbance due to the monolayer will be overwhelmed by the intense fringes. By using very thin mica, ~ 1 μm thick (Figure 2a), the fringe period has been increased to ~ 3400 cm^{-1} and thus appears as a gentle sloping baseline (inset a). This baseline can be easily removed by subtracting the spectrum of the pure mica substrate or by simple linear baseline correction.

An added benefit of using thin mica is that the mica is optically transparent down to 1200 cm^{-1} and at least partially transparent down to the spectrometer frequency limit at 200 cm^{-1}. This allows spectral subtraction routines to be employed below 1200 cm^{-1} for extraction of spectral information of the adsorbed species from the mica background (*11*). Secondly, the thin films of mica can be placed in series and this increases the amount of sample probed by the beam which has led to a direct improvement in the detection limit of monolayer and submonolayer coverages on mica surfaces.

The attenuation of the optical signal by a monolayer is small and the upper limit on S/N is usually determined by the dynamic range of the analog-to-digital converters (ADC) (*9*). A simple method to diminish the ADC limitation, and improve the S/N ratio would be to increase the amount of sample probed by the beam. Since the ADC peak-to-peak noise level for a single scan is ~10^{-3} absorbance it may be necessary to average several

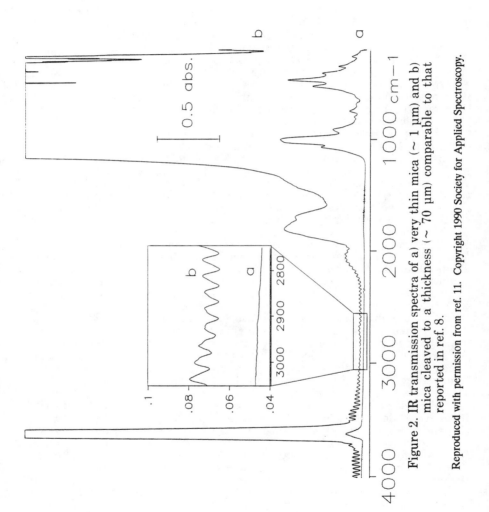

Figure 2. IR transmission spectra of a) very thin mica (~ 1 μm) and b) mica cleaved to a thickness (~ 70 μm) comparable to that reported in ref. 8.

thousand scans before the bands due to adsorbed species can be detected above the background noise. However, if a tenfold increase in the amount of sample was possible it would result in a one hundredfold decrease in the number of scans required or alternatively, for an equal number a scans, a 10 fold improvement in the detection limit. For monolayer studies an increase in the amount of sample can only be accomplished by probing a larger surface area. This would be difficult to achieve experimentally in reflection measurements because of geometric constraints.

For spectra of very thin mica recorded in transmission, an increase in surface area can be accomplished by simply placing several of the thin mica sheets in series. Although there is some increase in the adsorbance below 1300 cm^{-1} due to the mica, the fringing pattern should not change appreciably as long as the sheets of mica are approximately of the same thickness. The minimum distance between the sheets of mica is selected so that the secondary spike in the interferogram generated from reflection between the mica sheets appears outside the region of the interferogram used to compute the spectrum. Figure 3 shows the results of an experiment recorded with two thin mica substrates on which four monolayers of OTS were deposited. The Teflon holders could easily be stacked in series and anchored in position through a series of threaded connectors. A spacing of about 5 mm between the mica sheets was used. Curves a and b are the spectra of the two single mica substrates run separately while curve c is the result of placing both micas in series. Figure 4 is an enlargement of box a in Figure 3 and shows the C-H stretching region of OTS. The sloping baselines are due to the interference fringes. They are much broader than the infrared bands due to the absorbed monolayer and can be easily removed by subtracting a background spectrum of mica as shown in figure 4d.

The experiments using thin mica sheets have allowed us to easily measure the IR spectra of monolayers of materials such as OTS deposited by self-assembly and phospholipids deposited using a Langmuir trough. However, the extension of these measurements to the solid/liquid interface was not possible because of the delicacy of the thin films of mica. One solution to the problem of using thin mica at the solid/liquid interface was to employ an internal reflection element as a support for the mica sheet (*11*). The penetration of the IR beam into the rarer medium from the internal reflection element (IRE) is due to the presence of an evanescent wave, the intensity of which falls off exponentially with distance (*14*). For a ZnSe crystal operating at 45° the penetration depth is approximately 0.9 μm at 3000 cm^{-1} and 2.7 μm at 1000 cm^{-1}, sufficient to completely penetrate a very thin mica sheet in intimate contact with the crystal, and sample the interfacial region on the upper surface of the mica. Studies at the solid/liquid interface could be performed by placing the internal reflection element in a horizontal ATR accessory (Spectra-Tech 0012-303). The liquid cell consisted of a series of O-Rings held firmly against the surface of the mica not in contact with the IRE by pressure supplied by the gripper device of the ATR accessory. The liquid was injected into the centre of the O-rings through a slot in the top plate of the gripper device. Although we were able to record spectra in this manner we found it difficult to cleave very large sheets of thin mica of uniform thickness to cover the entire surface of the ATR crystal.

An alternative to supporting the mica on an internal reflection element is to use thicker mica sheets as windows in a liquid transmission cell. The thicker mica was required to prevent the deformation or tearing of the mica when liquids were injected into the cell, but this returned us to the

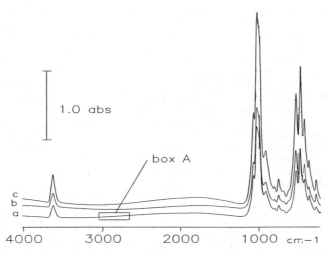

Figure 3. Spectra of OTS monolayer on two separate mica sheets. Both a) and b) were recorded separately and c) were a and b combined separated by 5 mm.

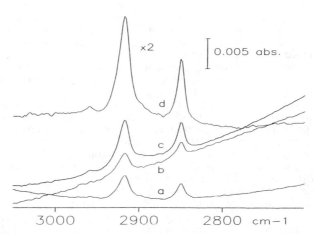

Figure 4. Expanded region of Box A in Figure 3. Figures a),b) and c) are defined in Figure 3 and d) is c) after subtraction of the background mica.

original problem of eliminating the fringes. However, the problem is not due to the existence of the fringes itself, but arises from the fact that we can not exactly reproduce the fringe pattern as a result of our inability to exactly reposition the mica in the spectrometer after deposition of the monolayer. For deposition of polymers or amphiphiles by self-assembly it should be possible to circumvent this problem by performing the adsorption of the monolayer *in situ*.

A liquid *in situ* transmission cell using mica for optical windows is shown in Figure 5. The front plate contained luer lock fittings for the introduction of liquid into and out of the cell. The front plate and back plate were cleaned of contaminants by heating in air at 600°C and then rinsed with filtered ethanol and blown dry with a stream of dry N_2. The mica windows were then glued to the front and back plates and were sufficiently thick to prevent deformation of the mica when the solvent was injected into the cell. For experiments involving organic solvents, the glue was glucose, and for adsorption from water, Epon 1004 was used. The back and front plates were placed into the holder, separated by a spacer of the appropriate thickness. A multiple film technique analogous to that described above for the thin film transmission experiments was possible and was used to acquire the spectrum of adsorbed OTS in the presence of CCl_4 (Figure 6). Twenty mica surfaces, (the inside surface of the two windows, plus nine mica sheets separated by Teflon spacers) were used in this experiment. The OTS had been deposited on the mica by the addition of 3mM OTS in cyclohexane to the cell, followed by the removal of the excess OTS/cyclohexane solution from the cell, rinsing with chloroform, and submerging in CCl_4. The spectrum was recorded in a single scan requiring 6 seconds to collect.

It was not necessary to use the multiple film technique to obtain spectra of adsorbed long chain molecules on mica as the intensity of the C-H asymmetric stretching modes (10^{-3} absorbance) was well above the detection limit of our FTIR spectrometer. Spectra of single sheets of mica with adequate S/N could be acquired with 15 minute collection time. For the adsorbed polymers on mica, however, the situation is more complex. For OTS, the amount adsorbed is about 3 mg/m^2 (eg., about 75 CH_2 groups/nm^2) whereas for PEO/PS block copolymers, typical adsorption values are about 1 mg/m^2, corresponding to only 6 CH_2 groups/nm^2. Therefore, for the strong CH asymmetric stretching vibration (assuming equal values for the absorption coefficients) there is order of magnitude difference in signal. Acquisition of spectra of equal quality to those obtained for OTS would require scanning 100 times longer (an unrealistic endeavor). In this case the multiple film technique is essential. Unfortunately, in many cases of interest to us the solvents used are aliphatic or aromatic hydrocarbons which have strong bands in the same spectral regions as the polymers. Rinsing the cell with an infrared transparent solvent such as CCl_4 (a good solvent for most of the polymers used) was avoided because of the possibility of removal of the polymer from the surface. Simply flushing the cell with dry N_2 was not effective because of the geometry of the multiple film cell.

Transmission–Reflection Studies

An alternative approach to the study of monolayers on mica was to use a reflection apparatus shown in Figure 7. The mica was placed in contact with a gold coated mirror and the infrared beam passed twice through the

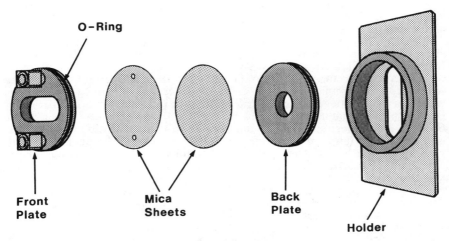

Figure 5. The *in situ* liquid cell.

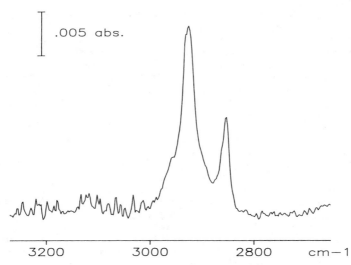

Figure 6. Spectrum of an OTS monolayer on mica recorded in the presence of CCl₄.

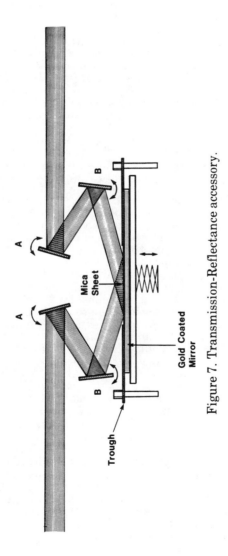

Figure 7. Transmission-Reflectance accessory.

sample. Variable low angle reflection could be performed using the optical path illustrated in Figure 7 whereas near normal reflection could be accomplished by using the mirrors labeled A only. When operated in the near normal configuration a high level of fringe suppression was achieved due to the fact that the fringe pattern from the reflected beam undergoes a 180 degree phase shift (*14*). Complete removal of the fringes was accomplished by performing the adsorption *in situ* or, as with the previously described transmission experiments, the fringe problem could be circumvented by using thin mica sheets. In the latter method, monolayers deposited from the Langmuir trough could be studied. Orientation measurements were possible by using polarizers and the angle of incidence could be easily changed by raising or lowering the gold coated mirror. Extension to studies at the solid/liquid interface was accomplished by gluing the mica to the bottom of a liquid cell, anchoring the cell to the mirror, and performing the experiment *in situ*. Removal of the excess solvent/adsorbate after adsorption was easily accomplished by a combination of aspiration and rinsing steps. We would aspirate 95% of the solvent/adsorbate solution, add pure solvent and repeat this step several times until the final quantity of adsorbate in solution was < 1% of the amount required for monolayer coverage. Final aspiration removed > 99% of the solvent/adsorbate solution in contact with the mica. The residual solvent was allowed to evaporate. In this way, the amount of adsorbate cast onto the surface of the mica was negligible.

An adsorption isotherm of OTS on mica was recorded in this manner and is shown in Figure 8. A 3mM solution of OTS in cyclohexane was added to the liquid cell and was withdrawn (via aspiration) at various time intervals, followed by several rinses of the mica with chloroform. The adsorption seems to be complete after 2 min; there is no evidence of multilayer formation or of OTS cast on the surface from any residual solution after aspiration. An area/molecule of 0.2 nm^2 was measured for a monolayer coverage of OTS on mica. The area/molecule was obtained from a Beer's Law relationship calculated by using the intensity of the band at 2920 cm^{-1} calibrated with standard solutions of OTS in CCl$_4$. This area/molecule is in agreement with those reported by Sagiv *et. al.* (*15*) for OTS on silicon and glass surfaces using conventional ATR techniques. OTS adsorption is insensitive to surface composition (*15*) or roughness (*16*) because of the formation of Si-O-Si linkages between OTS molecules.

It is difficult, if not impossible, to extend these reflection experiments to multiple films with this technique. Thus, to measure the spectra of adsorbed polymers on mica an improvement in the detection limit of our spectrometer was required. One method used for reducing the ADC noise limitation is to operate in an optical null mode (*17*). In this method the interferogram measured is due to the radiation absorbed by the sample. For weakly absorbing samples (eg. monolayers) the measured interferogram is small and can be amplified to fill the dynamic range of the ADC or until detector noise is detected. The improvement in S/N is directly related to the amount of amplification. We have built a dual beam optical null FTIR (unpublished data) specifically designed for monolayer measurements which has allowed us to record spectra of adsorbed polymers . An example of an adsorbed polymer spectrum recorded with this device is shown in Figure 9. By optimizing the source intensity, using appropriate optical filters and operating in an optical null mode we have recorded spectra which have a peak-to-peak noise level equivalent to 2x 10^{-5} absorbance in a sampling time of 1 second. This improved sensitivity has enabled us to

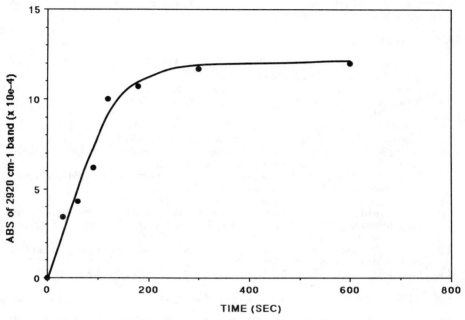

Figure 8. Adsorption Isotherm of OTS on mica.

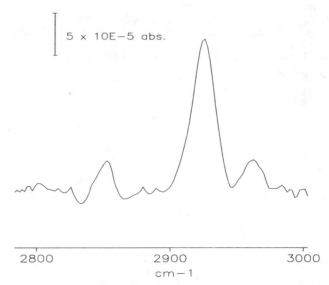

Figure 9. Spectrum of a monolayer of the copolymer PEO/PS on mica.
Spectrum was recorded with a sampling time of 1 second.

obtain spectra of sub-monolayer amounts of adsorbed polystyrene and to measure adsorption isotherms.

Literature Cited

1. Hurlbut, C. S. Jr.; Klein, C. *Manual of Mineralogy*; John Wiley and Sons: New York, 1977; p 391.
2. Poppa, H.; Elliot, A. G. *Surf. Sci.* **1971**, *24*, 149.
3. Israelachvili, J. N; Adams, G. E. *J. Chem. Soc.* Faraday Trans. 1. **1978**, *74*, 975.
4. Patel, S. S.; Tirrel, M. *An. Rev. Phys. Chem.* **1989**, *40*.
5. Marra, J.; Hair, M. L. *Colloids and Surfaces* **1988/89**, *34*, 215.
6. Hair, M. L. In *Vibrational Spectroscopies for Adsorbed Species*; Bell, A. T., Ed.; ACS Symposium Series No. 137, 1980.
7. Griffiths, P. R.; de Haseth, J. A. In *Fourier Transform Infrared Spectrometry*; J. Wiley and Sons: New York, 1986.
8. Carson, G.; Granick, S. *Appl. Spectrosc.* **1989**, *43*, 473.
9. Hirshfeld, T. In *Fourier Transform Infrared Spectroscopy: Applications to Chemical Systems*; Farraro, J. R.; Basila, L. J., Eds.; Academic Press: New York, 1979, Vol. 2, p 193–239.
10. *Optical Spectroscopy: Sampling Techniques Manual*; Harrick Scientific Corporation, 1987.
11. Guzonas, D. A.; Hair, M. L.; Tripp, C. P. *Appl. Spectrosc.* **1990**, *44*, 290.
12. Tlili, A; Smith, D. C.; Beny, J. M.; Boyer, H. *Mineral Mag.* **1989**, *53*, 165.
13. Langer,K.; Chatterjee, N. D.; Abraham, K. *Neues Jahrb. Mineral. Abh.* **1981**, *142*, 91
14. Harrick, N. J. In *Internal Reflection Spectroscopy*; Harrick Scientific Corporation, 1987; 3rd edition.
15. Maoz, R.; Sagiv, J. *J. Colloid Interface Sci.* **1984**, *100*, 465.
16. Pomerantz, M.; Segmuller, A.; Netzer, L.; Sagiv, J. *Thin Solid Films* **1985**, *132*, 153
17. Kuehl, D.; Griffiths, P. R. *Anal. Chem.* **1978**, *50*, 418.

RECEIVED August 2, 1990

Chapter 15

Ethoxylated Alcohols at Solid Hydrocarbon Surfaces

Detergency Mechanisms

David R. Scheuing

Clorox Technical Center, 7200 Johnson Drive, Pleasanton, CA 94588

Time - resolved spectra of a solid hydrocarbon layer on the surface of an internal reflection element, interacting with an aqueous solution of a nonionic surfactant, can be used to monitor the detergency process. Changes in the intensity and frequency of the CH_2 stretching bands, and the appearance of defect bands due to gauche conformers indicate penetration of surfactant into the hydrocarbon layer. Perturbation of the hydrocarbon crystal structure, followed by displacement of solid hydrocarbon from the IRE surface, are important aspects of solid soil removal. Surfactant bath temperature influences detergency through its effects on both the phase behavior of the surfactant solution and its penetration rate into the hydrocarbon layer.

The removal of liquid oily soils from surfaces is generally understood in terms of three basic mechanisms: the roll - back of droplets of oily soil, the surfaces of which are modified by the presence of an adsorbed layer of surfactant; direct emulsification of macroscopic droplets of soil; and the direct solubilization of the oily soil into surfactant micelles or other interfacial phases formed (1-3).

Solid soils are commonly encountered in hard surface cleaning and continue to become more important in home laundry conditions as wash temperatures decrease. The detergency process is complicated in the case of solid oily soils by the nature of the interfacial interactions of the surfactant solution and the solid soil. An initial soil softening or "liquefaction", due to penetration of surfactant and water molecules was proposed, based on gravimetric data (4). In our initial reports of the application of FT-IR to the study of solid soil detergency, we also found evidence of rapid surfactant penetration, which was correlated with successful detergency (5). In this chapter, we examine the detergency performance of several nonionic surfactants as a function of temperature and type of hydrocarbon "model soil". Performance characteristics are related to the interfacial phase behavior of the ternary surfactant - hydrocarbon - water system.

A correlation of the detergency performance and the equilibrium phase behavior of such ternary systems is expected, based on the results presented by Miller et al. (3,6). The phase behavior of surfactant - oil - water (brine) systems, particularly with regard to the formation of so-called "middle" or "microemulsion" phases, has been shown by Kahlweit et al. (7,8) to be understandable in terms of the

0097–6156/91/0447–0251$08.00/0

interplay of the water - oil and surfactant - water miscibility gaps. In studies of the detergency performance of nonionic surfactants for liquid soils such as hexadecane (C_{16}) and mineral oil, Miller et al. (3) showed that maximum rates of oil solubilization and removal from fabrics occurred near the "phase inversion temperature" (PIT) of the ternary surfactant - soil - water system. The PIT was defined as the temperature at which the system forms a middle phase microemulsion containing equal volumes of water and oil in equilibrium with excess water and oil phases. The low interfacial tensions at the PIT enhance fabric cleaning because of the higher efficiency of mechanical agitation processes in aiding spontaneous emulsification. Videomicroscopy also revealed that rapid solubilization of C_{16} into the L_α phase of the nonionic surfactant initially present at the oil-water interface at such temperatures can occur. As work on spontaneous emulsification has shown, diffusion path theory can predict conditions favorable to spontaneous emulsification, if the equilibrium phase diagram is known for the ternary system (9). Additional work on the kinetics of liquid soil removal has also indicated that the properties of the interfacial mesophases formed at the soil - water interface (viscosity, composition) can affect the overall efficiency of the detergency process (10). It should be noted that such detergency studies are focussed on what may be considered practical "worst case" systems of oily soils on relatively nonpolar substrates, where roll-up of the oily soil is not the primary removal mechanism.

In ellipsometric studies of triglyceride removal from plastic and metal surfaces (11), Lindman et al. concluded, for nonionic surfactants, that detergency performance was enhanced when the surfactant aggregate's curvature against water, soil, and substrate was as close to zero as possible. This observation indicated that selection of temperature and surfactant type can be optimized to provide a planar layer of surfactant molecules at the solid soil surface, which was again related to the phase behavior of the ternary surfactant - hydrocarbon - water systems. The effects of additives such as hydrocarbons and alcohols on the detergency performance of nonionics, all at relatively low temperature (25 °C), were consistent with the changes in the critical packing parameter (CPP) of the surfactant aggregates caused by the additives. Larger, nonspherical micelles of a nonionic surfactant such as $C_{12}EO_5$ (larger CPP) exhibited better detergency performance than smaller, spherical micelles of a surfactant such as $C_{12}EO_8$.

FT-IR , using attenuated total reflectance (ATR) sampling techniques, is quite suited for the study of the interfacial interactions of nonionic surfactants and solid soils. We can create a solid soil - solution interface by coating the internal reflection element (IRE) used in ATR with a layer of solid hydrocarbon. We then monitor surfactant adsorption by recording time - resolved spectra of the layer using software designed for on - the - fly data acquisition in gas chromatography - infrared spectroscopy (GC-IR). As is shown in a number of other chapters in this volume, FT-IR can provide information about the changes in the packing of long methylene chains, a common structural unit of solid oily soils. Thus, the perturbation of the structure of the solid soil by penetration of surfactant and water molecules, and the correlation of changes in the interfacial composition with detergency performance can be obtained.

Experimental

All spectra were obtained with a Digilab FTS 15/90 spectrometer, equipped with a liquid nitrogen cooled wide band mercury-cadmium-telluride detector. Digilab GC-IR data acquisition software (GC-32), running on a Model 3240 computer, was used to continuously record interferograms. Time - resolved spectra of nominal 4 cm^{-1} resolution were computed, using triangular apodization, from sets of 12 co-added interferograms. The effective time resolution was less than 0.1 minutes. Standard Digilab software was used for subtraction of the spectra of water (liquid and vapor) from the time - resolved spectra, and for plots of all spectra shown. Linear baselines have been subtracted from spectra to facilitate presentation, but no additional smoothing or noise reduction schemes were applied to any of the spectra used in this study. Band intensities and frequencies were determined from spectra receiving similar baseline corrections, using standard Digilab software routines.

In earlier studies (5,12), we coated a ZnSe IRE of a cylindrical internal reflectance cell (CIRCLE, Spectra-Tech Inc.) with a layer of model soil for interaction with commercial surfactant solutions. In this study we used a Harrick Prism Liquid Cell as the ATR sampling optics. This accessory features a rectangular IRE, in which the infrared beam undergoes nine internal reflections. The IRE can be dipped into a thermostatted beaker (10 ml) containing the surfactant solution. The IRE, in its holder, can be simply removed from the mirror assembly for coating with a layer of model soil. An advantage of the rectangular IRE of the Prism Cell is the retention of the polarization of the infrared beam, which is potentially useful in studies of orientation of adsorbed surfactants or of the soil itself. The CIRCLE, on the other hand, scrambles the polarization of the beam as it propagates down the length of the rod-shaped IRE.

The IRE of the Prism Cell was coated with a layer of hydrocarbon "model soil" in the same way as was done in earlier work with the CIRCLE (5,12). The IRE is withdrawn mechanically from the solution of the hydrocarbon of interest (2-4 wt.% in hexane) at a slow, controlled rate. The hexane flashes off, leaving a layer of hydrocarbon behind. By varying the withdrawal rate and the concentration of the hydrocarbon, layers of varying thicknesses, as judged by the intensity of the bands of the hydrocarbon spectrum ,can be obtained.

The sampling depth , d_p, of the infrared radiation from the surface of the IRE outward can be calculated from Equation 1 (13);

$$d_p = \frac{\lambda_1}{2\pi (\sin^2\Theta - n_{21}^2)^{1/2}}$$

where λ_1 is the wavelength of the radiation in the IRE (given by λ_1/n_1), Θ is the angle of incidence at the IRE-hydrocarbon interface, and n_{21} is the ratio of the refractive index of the hydrocarbon to that of the IRE. For a ZnSe IRE with $n_1 = 2.4$, a hydrocarbon with $n_2 = 1.5$, and a 45° angle of incidence, $d_p = 1.2$ µm at the frequency of the intense water bending band near 1640 cm^{-1}. The water bending band was always observed immediately upon immersion of the hydrocarbon coated IRE into the aqueous solutions, and thus the average thickness of the hydrocarbon layers must have been slightly less than d_p. It should be noted that the hydrocarbon layers generated in this manner are meant to model "real world" soil layers, and not molecular monolayers with any particular orientation.

The removal of hydrocarbon from the IRE surface can be monitored by the changes in the intensity of the intense CH$_2$ stretching band near 2850 cm^{-1} in the series of time-resolved spectra recorded during the exposure of the layer to surfactant solution. The absolute intensity of this band varies somewhat from layer to layer. Normalized intensities were obtained by dividing the intensity of the band in the spectrum of the initial, dry layer by the intensity of the band in each of the time-resolved spectra. These normalized intensities are plotted versus time. Values slightly greater than 1.0 occur because of the difference in refractive index between air and water, the media "behind" the thin hydrocarbon layers in the case of the initial and time - resolved spectra, respectively. Normalized intensities in excess of 1.0 can only be detected in detergency runs where little or no removal occurs.

A quantitative comparison of the kinetics of hydrocarbon removal based on the changes in band intensities was not attempted. Extracting quantitative hydrocarbon layer thicknesses from the band intensities is difficult, due to the change in the relative thickness of the optically distinct hydrocarbon and water "layers" within the sampling depth. The detergency process causes uncertainty in the value of the refractive index of the interfacial layer. Layer thicknesses can indeed be calculated from ATR spectra, as has been shown by Iwamoto et al. (14), and Sperline et al. (15), when the nature of the refractive index gradient at the IRE surface (in the case of "thick films") is known (14) or can be estimated in a reasonably simple manner (15,16).

Hydrocarbons (99 % purity) were from Aldrich and Sigma Corps. The nonionic surfactants used in this work were single component alkyl poly(ethyleneoxides) (for example $C_{12}EO_5$), obtained from Nikko Chemicals Co. (Japan), dissolved in HPLC grade water at a concentration of 0.03 wgt. %. The use of static surfactant solutions at this low concentration has been found (5) to slow the removal process enough to allow the acquisition of several time - resolved spectra within the first 3 minutes of exposure.

Screening Studies, Eicosane Model Soil

As a point of comparison with our earlier studies, eicosane (C_{20}) was used as the model soil in a series of screening experiments, employing the rectangular IRE of the Harrick Prism Liquid Cell as a substrate. The plots of the normalized intensity of the v_s CH_2 band as a function of time of exposure to various micellar solutions of monodisperse nonionic surfactants containing 6 EO groups are shown in Figure 1. Uniformly poor detergency performance is indicated by the small decreases in the band intensity at ambient temperature (25-27 °C). In the case of $C_{12}EO_6$, removal is not much improved by even a ten-fold increase in concentration.

Figure 2 indicates that, among the C_{12} types, the performance of $C_{12}EO_4$ was clearly superior to the more hydrophilic $C_{12}EO_6$ or more hydrophobic $C_{12}EO_3$. This observation lead to the detailed investigation of the temperature effects on the performance of $C_{12}EO_4$ discussed below.

Hexadecane Studies. Figure 2 also indicates that the removal of low viscosity liquid soils such as hexadecane (C_{16}) is difficult to monitor with this spectroscopic approach because of the rapidity of removal. The run performed with surfactant-free water, however, indicates that a stable C_{16}-water interface is achieved after 5 minutes. The somewhat hydrophobic ZnSe surface of the IRE can be considered a reasonable model of the polyester fabric surface. Water alone cannot completely clean the C_{16} from the surface, indicating a significant adhesion energy of hydrocarbon to the ZnSe.

The removal of C_{16} by the $C_{12}EO_4$ was extensive but incomplete. An examination of the spectra of the "residual" C_{16} layers is of value in interpretation of other spectroscopic changes in the solid hydrocarbons discussed below. Table 1 lists general band assignments for the spectra of the hydrocarbons and surfactants.

Figure 3 shows part of the so - called "fingerprint" region of the time - resolved spectra from the C_{16} runs. The frequency of the δ_s CH_3 band and the relative intensities of the CH_2 wagging defect bands indicate the presence of fully disordered methylene chains, as expected for a liquid hydrocarbon (17,18). Little change in the ordering of the chains occurs during thinning of the layer of C_{16} exposed to surfactant - free water. Lowering the temperature of the stable C_{16} layer, by chilling the water in contact with the layer, to 18 C , and then to 11 C, allows the spectroscopic changes associated with crystallization to be observed. As C_{16} freezes, the defect modes decrease in relative intensity, and the δ_s CH_3 band shifts toward lower frequency. At 18 °C, three ill-defined bands appear between 1260 and 1200 cm^{-1}. These bands decrease in intensity and narrow at 11 °C into members of the CH_2 "wagging band progression" (19). They may be assigned to CH_2 wagging

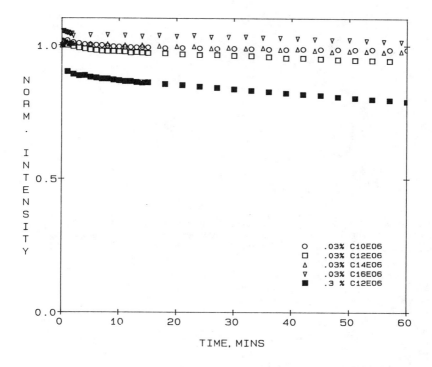

Figure 1. Screening study of removal of C_{20} from IRE surface by C_nEO_6 surfactants at ambient temperature. In this and other removal plots below, normalized intensity is that of the CH_2 symmetric stretching band of the hydrocarbon model soil.

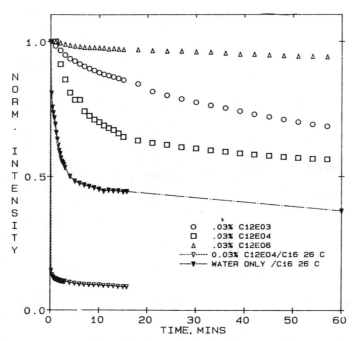

Figure 2. Comparison of C_{20} removal (ambient temperature) by $C_{12}EO_n$ surfactants (top three curves) and removal of liquid C_{16} by water only and $C_{12}EO_4$ (26 °C, bottom two curves).

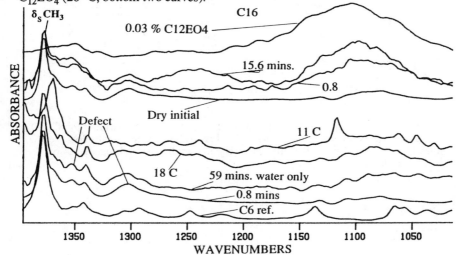

Figure 3. Normalized time - resolved spectra from C_{16} detergency runs. From bottom; liquid hexane reference; C_{16} exposed to surfactant-free water at 26 °C, then chilled to 18 and 11 °C; dry C_{16} layer; then exposed to 0.03% $C_{12}EO_4$ at 26 °C; $C_{12}EO_4$ reference solution.

Table 1. Infrared Band Assignments

Wavenumber	Assignment
2960	ν_{as} , CH_3 asymmetric stretch
2917	ν_{as} , CH_2 asymmetric stretch
2876	ν_s , CH_3 symmetric stretch
2850	ν_s , CH_2 symmetric stretch
1471	δ_s , CH_2 "scissoring " deformation
1456	δ_{as}, CH_3 asymmetric deformation
1378	δ_s , CH_3 symmetric deformation
1367, 1306	CH_2 wag , "defect mode" t (gtg') t "kink"
1342	CH_2 wag , "defect", (gt) terminal end - gauche
1354	CH_2 wag ,"defect", t (gg) t double gauche ,bent chain
1345	CH_2 wag of CH_2 - O - CH_2 , surfactants
1250, broad	CH_2 twist of CH_2 - O - CH_2 , surfactants
1150-1050	coupled C -O stretch, C-C stretch, CH_2 rock ,surfactants
1120,1070	C -C stretch, hydrocarbons 2 or 3 bands, exact frequency is dependent on chain length

of methylene groups adjoining certain gauche defects, as has been shown in studies of liquid hydrocarbons and polyethylene (18). A reference spectrum of liquid hexane is included in the figure, to further illustrate the CH_2 wagging bands of gauche conformers. In such a short hydrocarbon chain, fewer total gauche conformations are possible, which results in lowered intensity of the defect bands near 1354 and 1341 cm^{-1}, the splitting of the 1300 cm^{-1} band into a multiplet, and the appearance of more distinct bands near 1250 and 1220 cm^{-1}.

The time-resolved spectra from the run of C_{16} exposed to $C_{12}EO_4$ (Figure 3) are quite different from those of the surfactant - free water run. The relative intensity of the end gauche and double gauche defect bands is increased, and the kink defect band bear 1300 cm^{-1} may be split into a multiplet. In addition, several bands between 1260 and 1200 cm^{-1} are enhanced, relative to the spectrum of the initial, dry layer. The spectrum of a layer of neat $C_{12}EO_4$ clinging to the surface of the IRE (Figure 4) indicates that the surfactant can contribute bands near 1351, 1323, 1300, and 1250 cm^{-1}, but only at significant interfacial concentrations. In such cases very intense C-O bands of the surfactant are also observed. The reference spectrum of 0.03% $C_{12}EO_4$, recorded in the absence of a C_{16} layer, is plotted in Figure 3 to the same scale expansion as the time - resolved spectrum (at 15.6 minutes) from the $C_{16}/C_{12}EO_4$ detergency run. Since relatively weak C-O bands between 1150 and 1050 cm^{-1} are found in the spectra from the detergency run, the bands between 1260 and 1200 cm^{-1}, as well as the other changes, are interpreted as due to partial ordering of the methylene chains of the residual C_{16} under the influence of a small amount of adsorbed $C_{12}EO_4$. Lowering the temperature of the C_{16} layer to the freezing point in the presence of water seems to cause similar chain ordering. The residual C_{16} may not be completely liquid - like, when incorporated into a ternary hydrocarbon - surfactant - water phase.

The spectrum of a layer of a lamellar liquid crystal phase, composed of 46 % C_{16}, 31.6 % $C_{12}EO_4$, and 22.4 % water is shown in Figure 4. Spectroscopic changes occur upon exposure of this layer to water, as illustrated by several normalized time - resolved spectra. Uptake of water by the lamellar phase results in increased hydrogen bonding of the EO groups (5) detected by a very rapid shift of the surfactant C-O bands toward lower frequency. A comparison of the spectra in Figure 4 to those in Figure 3 indicates that the residual C_{16} layer contains a much lower concentration of $C_{12}EO_4$ than is found in either the lamellar liquid crystal phase or the more "hydrated" phase or mixture of phases resulting from the uptake of water by the liquid crystal.

Eicosane / $C_{12}EO_4$ Effect of Temperature on Removal

Figures 5 and 6 show the effect of temperature on the removal of solid C_{20} (melting point = 37 °C) by $C_{12}EO_4$. These plots of normalized intensity of the v_s CH_2 band versus time were obtained from two series of experiments, in which the initial layer thickness was varied somewhat. As discussed above, these plots must be regarded as qualitative descriptors of the removal process, due to the optical complexity of the interface. The removal process may involve not only solubilization, but also a surfactant - induced displacement of C_{20} crystallites from the IRE surface, which cannot be treated as a gradual thinning of the C_{20} layer. Repeated experiments on the effect of temperature on removal rate indicate that if the conditions of layer preparation (hydrocarbon concentration in hexane, speed of withdrawal from the solution) are held constant, then reproducible band intensities of the initial layers are obtained. The shape of the removal plots (Figures 5 and 6) are affected by the initial layer thicknesses. More rapid removal was usually observed for thinner layers of smaller initial C_{20} band intensity.

Removal of C_{20} by $C_{12}EO_4$ is somewhat faster, but still incomplete, at 26-27 °C, compared to that achieved at lower temperatures. As the temperature is raised above 28-30 °C, the initial apparent removal rates, as well as the extent of removal, seem to increase sharply. Unlike liquid C_{16}, essentially no removal of solid C_{20} occurs when the layer is exposed to surfactant - free water, even at 32 °C.

The temperatures of all these experiments are above the so- called cloud point of the $C_{12}EO_4$, which is about 4 °C (11). The solutions of the surfactant (0.03 weight %) were indeed visually hazy at room temperature.

$C_{12}EO_4$ undergoes a change near 18 °C from a two phase W+L_1 system to a W+L_α system. (6). The improvement in detergency performance of this W+L_α system over that of the micellar systems shown in Figure 1 is thus correlated with the appearance of lamellar liquid crystals of the surfactant in the bath. The temperature dependence of the performance of liquid crystals of $C_{12}EO_4$ between 22 and 32 °C, on the other hand, is probably due to changes in the nature of the

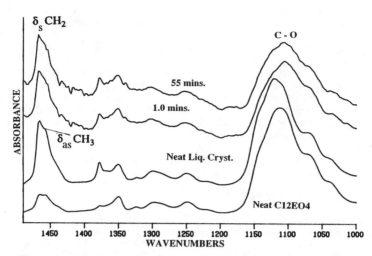

Figure 4. Reference and time - resolved spectra of neat $C_{12}EO_4$, and L_α phase of $C_{12}EO_4/C_{16}/H_2O$ water.

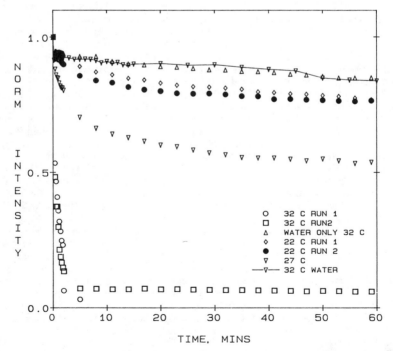

Figure 5. Temperature dependence of removal of C_{20} by $C_{12}EO_4$, and water only at 32 °C, "thicker" layers.

interfacial action of the liquid crystals upon contact with the solid hydrocarbon, as discussed further below in terms of the rate of penetration of the surfactant into the solid soil layer.

The importance of a surfactant - rich phase, particularly a lamellar one, to detergency performance was noted for liquid soils such as C_{16} and mineral oil (3,6). Videomicroscopy experiments indicated that middle phase microemulsion formation for $C_{12}EO_4$ and C_{16} was enhanced at 30 °C, while at 18 °C, oil - in - water, and at 40 °C, water - in - oil microemulsions were found to form at the oil - bath interface (3,6). A strong temperature dependence of liquid soil removal by lamellar liquid crystals, attributed to viscosity effects, has been noted for surfactant - soil systems where a middle - phase microemulsion was not formed (10).

An examination of the time - resolved spectra can be made, which provides additional details about the temperature - dependent changes in the nature of the hydrocarbon - bath interface during the detergency process.

Eicosane/$C_{12}EO_4$ Time - Resolved Spectra

Considering a solid hydrocarbon soil, changing the temperature of the washing bath will change several things simultaneously; the nature of the surfactant-rich phase ($W+L_1$ to $W+L_\alpha$), the viscosity of the L_α phase which has solubilized oil (10), the amount of surfactant adsorbed at the interface, and the "disorder" of the methylene chains of the hydrocarbon. The number of methylene chain defects in crystalline hydrocarbons increases with temperature, particularly near the chain ends (20). Spectroscopic studies have shown that the complex phase behavior of higher molecular weight (C_{19}-C_{45}) solid hydrocarbons can be linked directly to the effect of defect formation on lateral chain order (18). The crystal polymorphism of triglycerides is also known to be directly related to the methylene chain subcell packing adopted at a given temperature (21). As our earlier FT-IR studies of the removal of hydrocarbons (5) and triglycerides (22) have indicated, the penetration of nonionic surfactants into the layer of solid soil, and the subsequent "disordering" of the methylene chains is a step critical to the detergency process.

The effect of thermally induced disorder on the intense C-H stretching bands is presented first. A layer of C_{20} was prepared and spectra were obtained as the temperature of the surfactant - free water in contact with the layer was increased. As expected, a large shift toward higher frequency of the v_s and v_{as} CH_2 bands at the melting point was observed, because of the large increase in gauche conformers introduced by the change to completely disordered methylene chains (23). A decrease in relative intensity and loss of doubling of the v_{as} CH_3 band near 2960 cm^{-1}, due to the increased rotational motions of the methyl groups in the melt (24), was also found. The thermally induced changes in these bands below the melting point are more subtle, and are more easily studied with difference spectra (25,26), as shown in Figure 7.

The bottom series of difference spectra were obtained by subtraction of the spectrum of the C_{20} layer (in contact with water) at 20 °C from those obtained at higher temperatures. The changes in shape of the v_{as} CH_2 band with increasing temperature are complicated by the increases in intensity of the combination bands (19) near 2925 and 2905 cm^{-1}. The v_s CH_2 band broadens and increases in intensity relative to the v_{as} CH_2 band, in both the original and the difference spectra. The v_s CH_2 band does not exhibit a large increase in frequency until the melting point is reached. These changes in the v_s CH_2 band are probably due to a rearrangement or relaxation of some fraction of the C_{20} molecules in the layer caused by melting and recrystallization in the presence of a water interface. The molten C_{20} layer was then cooled to 20 °C and an additional spectrum recorded. The middle spectrum shows the differences caused by the melting process. The v_s CH_2 band is narrowed and shifted toward slightly lower frequency. The v_{as} CH_2 band is broadened toward

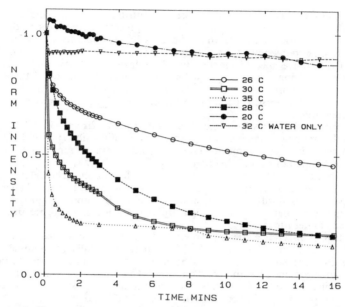

Figure 6. Temperature dependence of removal of C_{20} by $C_{12}EO_4$, "thinner" layers.

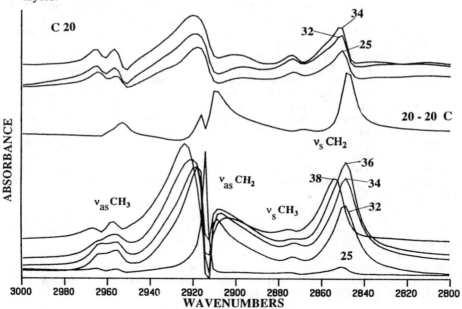

Figure 7. Normalized difference spectra, in C-H stretching region, of C_{20} layers exposed to water in temperature study. From bottom; differences in original layer with increasing temperature (spectrum at 20 °C subtracted); difference after recooling to 20 °C; differences caused by second heating.

lower frequency, probably due to increased contributions from the underlying combination band. A second heating of the layer produced the series of difference spectra at the top of the figure. Both of the CH_2 bands undergo a broadening and shift toward higher frequency as the temperature is increased a second time. These two series of difference spectra suggest that the relaxation process complicates the thermally-induced changes in the v_s CH_2 band somewhat, compared to those found in studies employing transmission (instead of ATR) optics. However, the band shifts observed are still of interest, especially in the case of C_{20} perturbations caused by surfactant penetration.

The normalized spectra from the detergency runs at 20 and 30 °C (Figure 8) indicate that the methylene chains of C_{20} remain in a solid - like state in both cases, even though there is a very large difference in the extent of removal by $C_{12}EO_4$ between these temperatures. If disordering of C_{20} by $C_{12}EO_4$ occurs, it is not as extensive as in the case of a true melting of the hydrocarbon. The relationship of subtle changes in these bands to the apparent removal rates can be investigated using difference spectra. Such difference spectra are obtained by subtracting the first time - resolved spectrum of the C_{20} layer (usually at 0.4 minutes exposure time) from those obtained at later times.

The series of difference spectra shown in Figure 9 were obtained from a run of C_{20} exposed to surfactant - free water at 32 °C, where essentially no removal occurs. The changes in these spectra must thus be due to slight thermal disordering and possibly a rearrangement of some fraction of the C_{20} chains on the surface of the IRE. Broadening of both CH_2 bands over 60 minutes exposure results in slow increases in components at lower frequency, near 2907 and 2845 cm^{-1}. At long exposure times, the contributions to the difference spectra from the combination modes near 2926 and 2890 cm^{-1} increase. The packing of the terminal methyl groups is apparently changed somewhat as well, as indicated by the bands near 2869 cm^{-1} (v_s CH_3, originally at 2871 cm^{-1}), and a doublet at 2859 and 2851 cm^{-1} (v_{as} CH_3, originally at 2861.8 and 2952.9 cm^{-1}). Additional bands in the fingerprint region will be discussed further below.

An initial period of rapid, but only partial removal occurs at 27 °C in the presence of $C_{12}EO_4$. Difference spectra from this run are shown in Figure 10. The intensity of the difference bands continues to increase throughout the entire 60 minutes of exposure. After 5 minutes, the difference spectra exhibit a change in band shape. The CH_2 bands increase in intensity, shift toward slightly lower frequency, and continue to narrow between 5 and 59 minutes. The relative intensity of the combination mode near 2890 cm^{-1} is increased at later times, and a broad, complex v_{as} CH_3 band is produced. All of these changes suggest that penetration of $C_{12}EO_4$ in the first several minutes of exposure causes disordering of a fraction of the methylene chains of C_{20}. At longer times, however, the slower, continuous changes in chain packing indicated by the CH_2 bands must indicate a slower rate of modification of C_{20} crystal structure.

The difference spectra in Figure 11 are obtained from the runs at 30 and 35 °C. The bandshape changes at 30 °C resemble those found at 27 °C, but occur much more rapidly. Even more extensive disordering of C_{20} occurs at 35 °C, as indicated by the broadened CH_2 bands and the broad v_{as} CH_3 singlet.

Changes in the fingerprint region of the spectra can also be related to surfactant penetration. The frequency and shape of the δ_s CH_2 band are very sensitive to intrachain interactions The CH_2 wagging defect modes and the C-O bands of the surfactant appear in this region as well.

Eicosane / $C_{12}EO_4$ Spectra Fingerprint Region. The very small shift in the δ_s CH_2 band in the time - resolved spectra shown in Figure 12 indicates that exposure of a C_{20} layer to surfactant-free water at 32 °C results in only very slight changes in the packing of the methylene chains. Complete disordering of the methylene chains

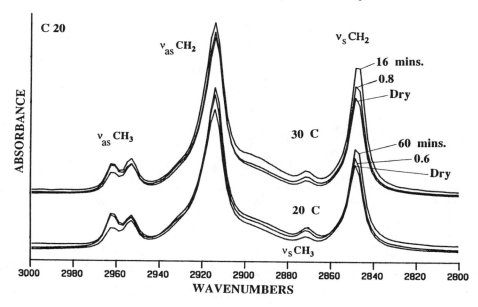

Figure 8. Normalized time - resolved spectra of C_{20} layers exposed to $C_{12}EO_4$ at 20 and 30 °C.

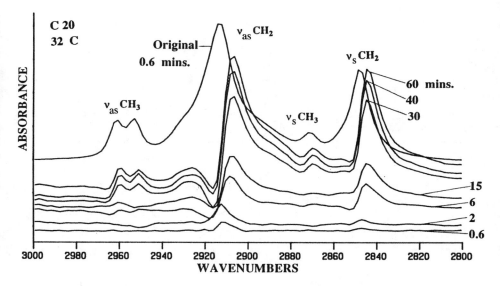

Figure 9. C_{20} layer exposed to water only at 32 °C. From bottom; time - resolved difference spectra (0.4 mins. subtracted); top = original spectrum at 0.6 minutes. Difference spectra all to same scale, expanded 15 X relative to original.

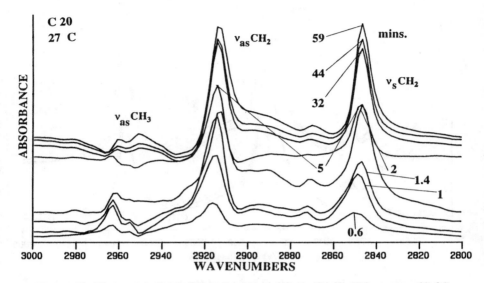

Figure 10. Time - resolved difference spectra from $C_{20}/C_{12}EO_4$ run at 27 °C. Bottom four spectra expanded 2X relative to top four spectra.

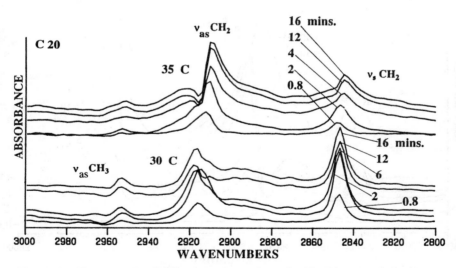

Figure 11. Time - resolved difference spectra from $C_{20}/C_{12}EO_4$ runs; 0.6 minute spectrum subtracted. Bottom five spectra from 30 °C run; top spectra from 35 °C run. All to same scale.

in the liquid state results in a shift toward lower frequency and a broadening of the δ_s CH_2 band in spectra of thin C_{20} layers, just as has been observed in a wide variety of studies of hydrocarbons (17), surfactants (27 -30), and phospholipid bilayers (23). The normalized spectra from the detergency runs at 27, 32 and 35 °C indicate a slight disordering of the C_{20}. However, in agreement with the changes exhibited by the C -H stretching bands, the disorder induced by penetration of $C_{12}EO_4$ is apparently not completely "liquid - like".

The spectra in Figures 13 and 14 illustrate other spectroscopic changes which occur during the interaction of $C_{12}EO_4$ and C_{20}. In the case of exposure of C_{20} to surfactant - free water at 32 °C, the lack of significant removal allows detection of the very weak CH_2 wagging and twisting bands. The absence of significant changes in the wagging/twisting bands, and the absence of significant defect bands indicate little change in the disorder of the C_{20} chains.

The rapid increase in the relative intensity of the defect modes is readily apparent in the case of successful detergency performance of $C_{12}EO_4$ at higher temperatures. Defect formation is found to be more rapid as the temperature is increased from 27 to 32 °C. As a comparison with the spectrum of molten C_{20} indicates, however, the disorder induced by $C_{12}EO_4$ is not completely "liquid - like". For example, the relative intensity of the broad "kink" band near 1300 cm^{-1} is smaller in the time - resolved detergency spectra than in the spectrum of the molten C_{20}. The bands between 1260 and 1150 cm^{-1} in the spectra from the runs at 30 and 32 °C appear to be related to the wagging/twisting progression. They resemble those bands produced in the spectra of C_{16} by lowering the temperature just to the freezing point, and in the spectra of the residual C_{16} interacting with $C_{12}EO_4$ at 26 °C. These bands are probably assignable to CH_2 wagging of specific gauche conformers of C_{20} formed by interaction with $C_{12}EO_4$. The broad, weak absorption between 1150 and 1050 cm^{-1} is due to C-O stretching of the $C_{12}EO_4$.

Visual observation of small flakes or sheets of solid C_{20} floating on the surface of the surfactant bath at the end of detergency runs at higher temperature indicated that a displacement mechanism is important to the very rapid removal monitored spectroscopically. The spectroscopic changes indicate that complete solubilization of C_{20} (resulting in liquid - like disorder in the hydrocarbon) is not required for removal. Penetration of a relatively small amount of $C_{12}EO_4$ is apparently capable of altering the adhesion of the solid C_{20} to the IRE surface.

The primary effect of temperature on the detergency performance of $C_{12}EO_4$ toward C_{20} is one of acceleration of penetration of the surfactant into the hydrocarbon layer. Increasing the temperature probably also causes a small increase in the CPP of the $C_{12}EO_4$ molecules involved in the formation of a relatively small amount of an intermediate phase at the C_{20} surface. Subsequent penetration of surfactant and water into C_{20} causes a rapid loss of adhesion of the hydrocarbon to the IRE.

By varying the hydrocarbon chain length, the relative importance of the chain disorder induced thermally and by penetration of $C_{12}EO_4$ can be investigated further.

Nonadecane Removal

Nonadecane, (C_{19}) like other odd carbon number hydrocarbons C9 to C45, undergoes a solid - solid phase transition below its melting point. For C_{19}, this change occurs at 22.8 °C, while melting occurs at 32 °C (18). The subcell packing of the methylene chains is orthorhombic below the transition temperature. The chains are fully extended in an all trans conformation with the planes of the carbon backbones of neighboring chains oriented at right angles. Above the transition, the subcell packing is "hexagonal", in the so - called "rotator" phase. The chains in this phase contain higher concentrations of non - planar conformers, which results in

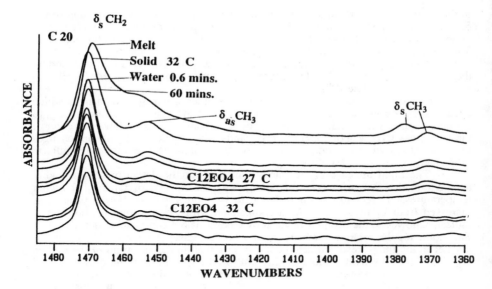

Figure 12. Time-resolved spectra from $C_{20}/C_{12}EO_4$ and C_{20}/water only runs. From top; molten and solid C_{20} references; exposed to water only (32 °C); $C_{12}EO_4$ (27 °C) at 0.4, 5, 60 mins.; $C_{12}EO_4$ (35 °C) at 0.8, 2, 16 mins. Scale expansions vary.

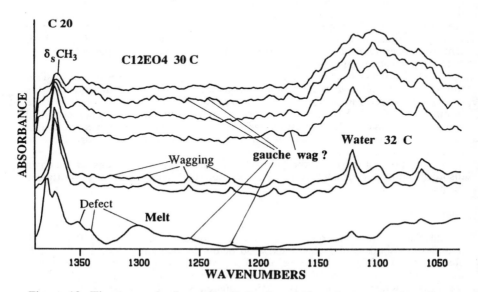

Figure 13. Time - resolved spectra of C_{20} layer. From bottom; molten C_{20} reference; exposed to water only (32 °C) at 0.6, 60 mins.; $C_{12}EO_4$ (30 °C) at 2, 8, 25, 60 mins. Scale expansions vary.

increased lattice dimensions and decreased interactions between neighbors. The orthorhombic to hexagonal transition of C_{19} is readily detectable spectroscopically by the collapse of the "factor group split" (31) δ_s CH_2 doublet (near 1472 and 1466 cm^{-1}) to a broader singlet. Above the melting point, the δ_s CH_2 band broadens considerably, shifts toward lower frequency, and decreases in intensity relative to the δ_{as} CH_3 shoulder near 1460 cm^{-1}.

Using a surfactant bath temperature of 19 °C initiates the transition of C_{19} from the hexagonal phase to the more ordered orthorhombic phase immediately upon starting the detergency experiment. The presence of $C_{12}EO_4$ does not prevent this transition, as indicated by the series of time-resolved spectra in Figure 15. The competing effects of increases in chain ordering and penetration of surfactant (leading to disordering and displacement from the IRE surface) result in lower reproducibility in removal of C_{19}. The normalized intensity of the v_s CH_2 band shown in Figure 16 indicates sluggish removal during the first several minutes of exposure at 19 °C, during the phase change. A period of rapid displacement between 2 and 5 minutes follows, after which removal slows again.

At temperatures above 22.8 °C, the $C_{12}EO_4$ interacts with the hexagonal phase of C_{19} directly. The normalized intensity plots exhibit smoother changes, and an increase in the apparent removal rate as the temperature is increased. The temperature dependence of the removal of C_{19} is not as sharp as with C_{20} or C_{26}.

The time - resolved spectra from the interaction of $C_{12}EO_4$ with hexagonal C_{19} indicate that penetration of the surfactant induces disordering of the C_{19} in a manner similar to that found for C_{20}. The δ_s CH_2 band in the spectra from the run at 30 °C shows that the methylene chains do not melt completely. In the 1400 - 1000 cm^{-1} region, (Figure 17) penetration of a small amount of $C_{12}EO_4$ is indicated by the increases in defect band intensities and the broad gauche defect bands between 1260 and 1150 cm^{-1}. These changes are more rapid and more extensive at 30 °C than at 26 °C.

The "looser" packing of the C_{19} methylene chains in the hexagonal phase can accommodate higher concentrations of non-planar conformers than is possible in the triclinic subcell of C_{20}. Intercalation of $C_{12}EO_4$ into the C_{19} layer occurs readily, resulting in the rapid formation of ternary phases at the IRE surface, and improved removal rates. Similar behavior was found for ethoxylated alcohols interacting with solid tristearin (22). In that FT-IR study, surfactant penetration into the α phase of tristearin (hexagonal subcell packing of methylene packing of chains) was found to be much more rapid than in the case of β' tristearin (triclinic subcell packing).

At temperatures just below the transition of C_{19} to the orthorhombic phase, penetration of $C_{12}EO_4$ is inhibited. However, the presence of the surfactant during the transition still affects the adhesion of the C_{19} crystallites to the IRE surface, resulting in some removal by the displacement mechanism.

Figure 18 shows the difference spectra in the C-H stretching region from the time-resolved spectra of the 30 °C run. The width of the CH_2 bands in these spectra increases significantly between 2 and 4 minutes, during the period of rapid removal. These spectra indicate the relatively greater disordering accomplished at this temperature, relative to the case of C_{20}. These difference spectra are thus consistent with the interpretation that surfactant penetration is more readily accomplished in the case of a hexagonal hydrocarbon phase.

The nature of the nonionic surfactant is still an important parameter in the removal of solid C_{19}. $C_{12}EO_5$ and the commercial surfactant Neodol 23-6.5 both form micellar solutions (L_1 phase only) at 25 °C. Detergency performance of these surfactants is poor. The presence of $C_{12}EO_4$ as a two phase system containing a surfactant - rich phase (L_1 or L_α) is still clearly beneficial to the kinetics of solid C_{19} removal.

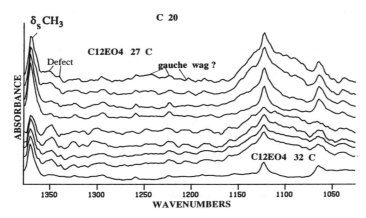

Figure 14. Time - resolved spectra of C_{20} layer exposed to $C_{12}EO_4$. From bottom; original, dry layer; 32 °C at 0.4, 0.6, 0.8, 1.2 mins.; 27 °C at 0.4, 5, 44, 59 mins. Scale expansions vary.

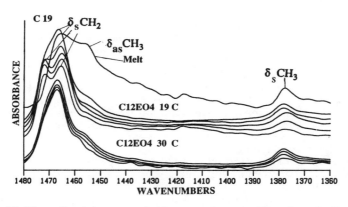

Figure 15. Normalized time - resolved spectra from $C_{19}/C_{12}EO_4$ runs. From top; molten C_{19} reference; exposed to $C_{12}EO_4$ (19 °C) at 0.7, 1, 1.8, 4, 7 mins.; $C_{12}EO_4$ (30 °C) at 0.2, 2, 20, 60 mins.

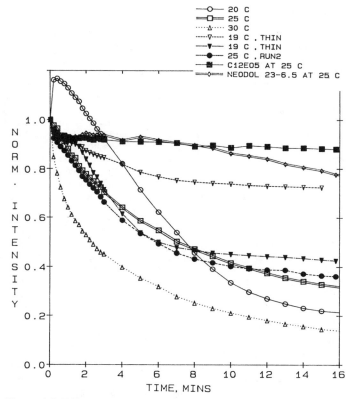

Figure 16. Effect of temperature on the removal of C_{19} by $C_{12}EO_4$.

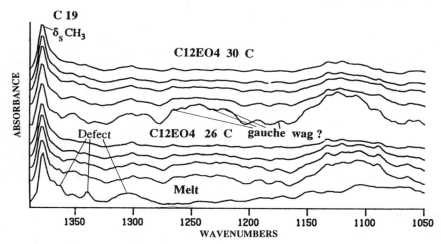

Figure 17. Normalized time - resolved spectra from $C_{19}/C_{12}EO_4$ runs. From top; exposed to $C_{12}EO_4$ (30 °C) at 0.2, 1, 4, 10, 60 mins.; $C_{12}EO_4$ (26 °C) at 0.2, 12, 32, 60 mins.; molten C_{19} ref.

Hexacosane Removal

The results obtained with C_{19} and C_{20} as model soils suggested that the crystal form of the hydrocarbon can affect surfactant penetration and hence removal. Hexacosane (C_{26}), with a melting point of 57 °C, allows investigation of the effect of temperature over a wider range than the systems described above. Additional details about the relationship of surfactant phase behavior to solid soil removal can be obtained, and the efficiency of the displacement mechanism can be explored further, using C_{26} as a model soil.

The normalized intensity plots in Figures 19 and 20 represent two series of experiments using C_{26} layers of different initial thicknesses. A significant increase in removal by $C_{12}EO_4$ is found at 29 -32 °C, whereas at 25-27 °C, the removal process slows after 5 minutes, with a short initial period of successful removal. Increasing the temperature above 32 °C results in slight increases in the apparent rate of removal during the first several minutes. Rapid displacement of C_{26} crystals at the higher temperatures was visually noted, as was found with the C_{19} and C_{20} layers. Examination of the time - resolved spectra in the fingerprint region confirmed that the interfacial behavior of $C_{12}EO_4$ and C_{26} was very similar to that of $C_{12}EO_4$ and C_{20}. Increasing temperature caused more rapid development of CH_2 defect bands and the broad bands between 1250 and 1200 cm^{-1}. The penetration of $C_{12}EO_4$ into the C_{26} layers is apparently aided by increases in temperature.

An interesting change in the time-resolved spectra from the detergency run at 52 °C can be detected. The normalized spectra from the runs at 32 and 52 °C are compared in Figure 21. The v_s CH_2 band undergoes changes at 32 °C which may be related to a relaxation or rearrangement of some fraction of the C_{26} methylene chains during the rapid penetration and removal by surfactant. The surfactant - induced changes in the CH_2 bands at 52 °C are somewhat different. A shift of v_s CH_2 from 2848.8 cm^{-1} to 2850.6 cm^{-1} and less change in the relative intensity of the v_s and v_{as} CH_2 bands are noted. The rapid collapse of the v_{as} CH_3 doublet into a singlet confirms a large increase in the rotational freedom of the methyl groups. These changes, as well as those of the δ_s CH_2 band shown in Figure 22, indicate that the disordered C_{26} is not as rapidly displaced from the IRE surface as at lower temperatures. A build - up of a ternary hydrocarbon - surfactant - water phase at the interface occurs.

The detection of disordered C_{26} at the interface at 52 °C might be due to the increased capacity of longer hydrocarbon chains for incorporation of non-planar "defect" structures in the solid state (18). Viewed from a detergency performance standpoint, the results suggest a change in mechanism with this hydrocarbon - "soil" combination at this relatively high temperature. Although displacement of much of the C_{26} occurs very rapidly, the relative importance of the displacement and solubilization mechanisms may change as the hydrocarbon soil increases in molecular weight. The detection of the disordered chains at 52 °C is consistent with more extensive solubilization of C_{26}. A ternary C_{26} - $C_{12}EO_4$ - water phase may be quite viscous and slow to diffuse away from the IRE surface, and thus enhance detection of disordered C_{26}. We are currently investigating the detergency performance of $C_{12}EO_4$, at 25 - 35 °C, against petrolatum as a soil. The performance of $C_{12}EO_4$ for this semi - solid, high molecular weight hydrocarbon mixture is indeed much poorer than that found in the case of C_{26}. Clearly, more work on a variety of hydrocarbon soils can be done to clarify these issues.

Hexacosane / $C_{12}EO_5, C_{12}EO_3$

The importance of the presence of a surfactant - rich phase in the washing bath has been mentioned several times. The performance of $C_{12}EO_5$ in removal of C_{26} is very poor at 30 °C, (Figure 23) just below the transition of the surfactant from a

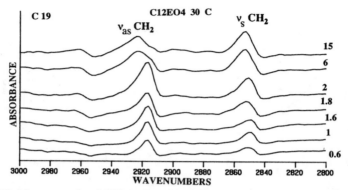

Figure 18. Time - resolved difference spectra from $C_{19}/C_{12}EO_4$ run at 30 °C, 0.4 minute spectrum subtracted. All to same scale.

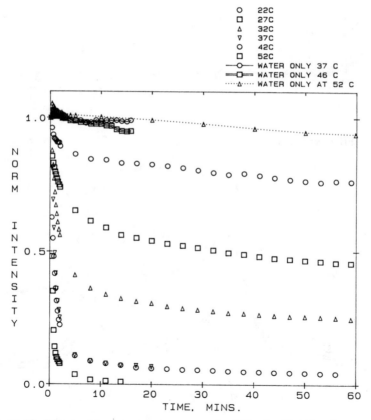

Figure 19. Effect of temperature on removal of C_{26} by $C_{12}EO_4$, first series of runs.

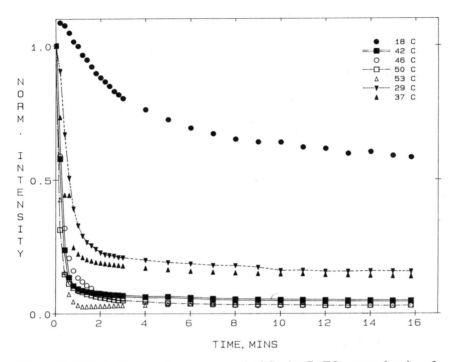

Figure 20. Effect of temperature on removal of C_{26} by $C_{12}EO_4$, second series of runs.

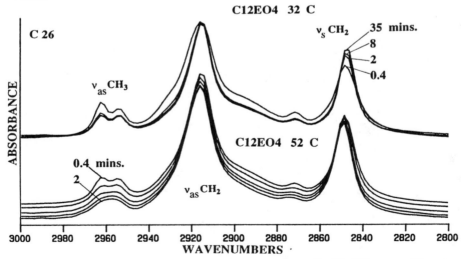

Figure 21. Normalized time - resolved spectra from $C_{26}/C_{12}EO_4$ runs. Top spectra, 32 °C run; Bottom spectra, 52 °C run at 0.4, 0.8, 1.2, 1.6, 2.0 mins.

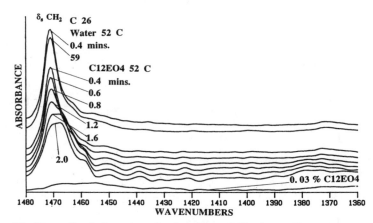

Figure 22. Normalized time - resolved spectra of C_{26} layer. From top; exposed to water only (52 °C); to $C_{12}EO_4$ (52 °C); Bottom spectrum is 0.03 % $C_{12}EO_4$ reference solution recorded with clean IRE, expanded to same scale as 2 minute spectrum from detergency run.

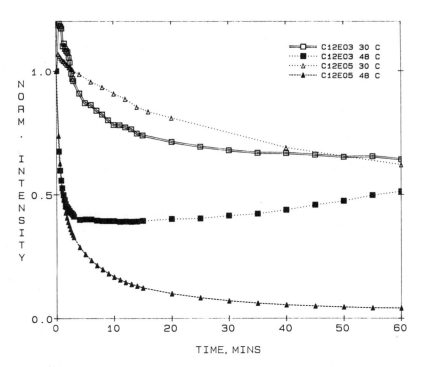

Figure 23. Effect of temperature on removal of C_{26} by $C_{12}EO_3$ and $C_{12}EO_5$.

single L_1 phase (micellar solution) to the $W+L_1$ phase at the cloud point of about 34 °C (11). The performance is improved considerably by raising the bath temperature to 48 °C, well above the cloud point but still well below the melting point of C_{26}. Visual displacement of crystals of C_{26} was observed at 48 °C, but not at 30 °C. An examination of the time - resolved spectra confirmed rapid penetration by the surfactant at 48 °C, by the changes in defect modes and the rapid increases in the difference spectra in the C-H stretching region.

$C_{12}EO_3$ is present as a $W+L_\alpha$ dispersion between 0 and about 30 °C (it does not exhibit a cloud point), and undergoes a transition to a $W+L_2$ system above 30 °C. A comparison of the detergency performance of the lamellar phases of $C_{12}EO_3$ and $C_{12}EO_4$ can be made at 30 °C. At 48 °C, the performance of the very hydrophobic L_2 phase can be compared with that of the L_α phase of $C_{12}EO_4$ and the L_1 phase of $C_{12}EO_5$.

The normalized intensity plot in Figure 23 indicates that the L_α phase of $C_{12}EO_3$ performs poorly at 30 °C. The difference spectra in Figure 24 show that the packing of the C_{26} methylene chains is rapidly perturbed by penetration of surfactant. At times greater than about 7 minutes, however, high frequency shoulders near 2930 and 2856 cm^{-1} appear on the CH_2 stretching bands. These highly shifted shoulders are due to liquid - like methylene chains. The v_{as} CH_3 doublet persists in the difference spectra at 59 minutes, rather than collapsing to a singlet, which is inconsistent with extensive hydrocarbon disordering. The appearance of the bands near 2930 and 2856 cm^{-1} in the difference spectra may thus be at least partially attributed to the presence of a significant amount of interfacial $C_{12}EO_3$. An examination of the time - resolved spectra in the fingerprint region (Figure 25) confirms that the interfacial concentration of $C_{12}EO_3$ at 30 °C is higher than that found in the cases of $C_{12}EO_4$ or $C_{12}EO_5$. The intensity of the broad C-O band due to surfactant increases steadily during the detergency run, eventually exceeding that found in a reference spectrum of 0.03 % $C_{12}EO_3$ recorded with a clean IRE. The increases in the defect modes of the hydrocarbon, which occur rapidly, suggest penetration by $C_{12}EO_3$. However, the relatively small amount of removal of C_{26} results in a build - up of an interacting surfactant-hydrocarbon phase at the IRE surface. Most of the bands in the spectra indicate that the interfacial phase is quite different from that established when lamellar $C_{12}EO_4$ contacts C_{26}. The relative intensity of the gauche wagging bands between 1260 and 1200 cm^{-1}, as well as those of the other CH_2 defect modes do not change much after the first several minutes of exposure. We suggest that the difference in performance of the lamellar phases of $C_{12}EO_4$ and $C_{12}EO_3$ is due to the dehydration of the $C_{12}EO_3$ upon penetration into the C_{26} layer, as was postulated in our earlier studies of the performance differences between two commercial nonionic surfactants (5).

At 48 °C, $C_{12}EO_3$ is present as a two-phase $W+L_2$ system. The surfactant - rich L_2 phase is very hydrophobic, and thus a phase separation of the surfactant onto the C_{26} layer occurs. The large increases in intensity of all bands due to $C_{12}EO_3$ in the time - resolved spectra allow ready detection of this phase separation.

Figure 26 shows the C-H stretching region of several time-resolved spectra from the run at 48 °C. The partial removal of C_{26} is indicated by the decrease in the C-H band intensity between 0.4 and 10 minutes. At longer exposure times, the drastic shift in the C-H bands toward higher frequency, and an actual small increase in overall intensity indicates the presence of a large amount of $C_{12}EO_3$ at the IRE surface. Reference spectra of the $C_{12}EO_3$ solutions at 30 and 48 °C, recorded with a clean IRE, are also shown. Interference of the C_{26} bands by bands due to adsorbed surfactant does not occur for the $W+L_\alpha$ solutions of $C_{12}EO_3$ at 30 °C (nor was any interference noted in the experiments with $C_{12}EO_4$ or $C_{12}EO_5$). The phase separation of $C_{12}EO_3$ at 48 °C, however, results in intense C-H bands due to surfactant even in the absence of the C_{26} layer. The use of the v_s CH_2 band intensity as a monitor of removal of C_{26} thus fails in the case of such large increases

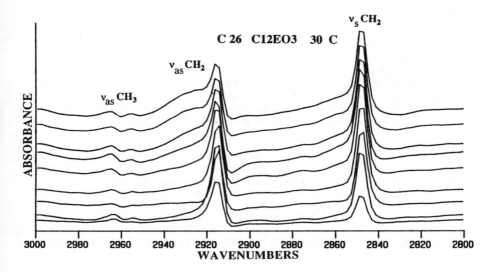

Figure 24. Time - resolved difference spectra from $C_{26}/C_{12}EO_3$ run at 30 °C, 0.4 minute spectrum subtracted. From bottom, time = 1, 2, 3, 4, 7, 14, 30, 45, 60 minutes. All to same scale.

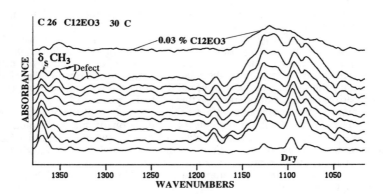

Figure 25. Time - resolved spectra from $C_{26}/C_{12}EO_3$ run at 30 °C. From bottom, original dry layer, 0.4, 2, 3, 4, 7, 14, 30, 60 minutes.; Top spectrum = 0.03 % $C_{12}EO_3$ reference at 30 °C recorded with clean IRE. All to same scale.

in the intensity of bands due to surfactant. Significant C-H bands due to C_{26} are detected even after rinsing the layer in water at 48 °C for 15 minutes. Clearly, the removal of C_{26} is incomplete, even after the establishment of the very high interfacial concentration of $C_{12}EO_3$. Such behavior is similar to that exhibited by a commercial surfactant, Neodol 25-3, studied earlier (5).

The fingerprint region of the time - resolved spectra (Figure 27) also indicate the huge increase in $C_{12}EO_3$ at the C_{26} - water interface due to phase separation. Complete displacement of C_{26} by the L_2 phase of $C_{12}EO_3$ does not occur. Thus, although penetration of this more hydrophobic surfactant can occur, a lack of involvement of water in the process results in poorer performance than can be accomplished at lower temperatures by the more hydrophilic $C_{12}EO_4$.

Conclusions

Solid hydrocarbon soils can be rapidly removed from the surface of a ZnSe IRE by alkyl poly(ethylene oxide) surfactants. The removal mechanism involves penetration of a small amount of the surfactant into the hydrocarbon layer, which causes an increase in methylene chain defects in the soil, and displacement of solid soil from the substrate. Solubilization of a large fraction of the solid soil is not required.

The presence of the nonionic as a two-phase system, either $W+L_1$ or $W+L_\alpha$, is beneficial to solid soil removal. The effect of temperature on detergency performance thus enters as an important variable, determining the surfactant phases initially present in the washing bath. At a given temperature, the ethylene oxide content of the surfactant is also an important variable, linked to the phase behavior. Solid soil removal by $C_{12}EO_5$ or $C_{12}EO_6$, which are present as micellar solutions (L_1 phase only) at "low" temperatures (20-30 °C) is poor. The time - resolved spectra obtained show a lack of evidence of hydrocarbon penetration by these surfactants. The lamellar phase of $C_{12}EO_4$ shows excellent detergency performance in the same temperature range, which can be linked to very rapid perturbation of the crystal structure of the hydrocarbon soil by surfactant penetration. The performance of the lamellar phase of $C_{12}EO_3$ is poorer than that of $C_{12}EO_4$ because of the tendency of the surfactant to dehydrate upon penetration into the hydrocarbon. The phase separation of the L_2 phase of $C_{12}EO_3$ onto the hydrocarbon layer can be readily detected spectroscopically by the relatively larger intensities of all surfactant bands. Partial penetration and removal of solid soil by the L_2 phase occurs, but the detergency performance is still poorer than that of the L_α phase of $C_{12}EO_4$.

The nature of the hydrocarbon affects the performance of the $W+L_\alpha$ system of $C_{12}EO_4$ through its effect on the rate of penetration of the surfactant. The "looser" chain packing of the hexagonal phase of C_{19} allows more rapid surfactant penetration than in the case of the triclinic crystals of C_{20} and C_{26}. A slightly higher temperature is required for rapid penetration of $C_{12}EO_4$ into C_{26}, compared to C_{20}, probably because of the greater accommodation of chain defects possible in higher molecular weight hydrocarbons.

The general trends in the detergency performance of the nonionic surfactants as a function of temperature and ethylene oxide content resemble those of the case of liquid hydrocarbon "model soils". A PIT, of course, is not strictly defined in the case of solid hydrocarbons. However, the penetration of a nonionic into the solid hydrocarbon can be likened to the partitioning of the surfactant between the water and liquid oil phases, which, when balanced at the PIT, results in excellent detergency performance. Unlike liquid oily soils, however, increasing temperature will increase solid soil removal by a properly selected nonionic surfactant because displacement of solid hydrocarbon, rather than complete solubilization, is a major part of the removal mechanism. Solid soil detergency is complicated by the melting of the hydrocarbon, which then changes the removal mechanism to one dominated

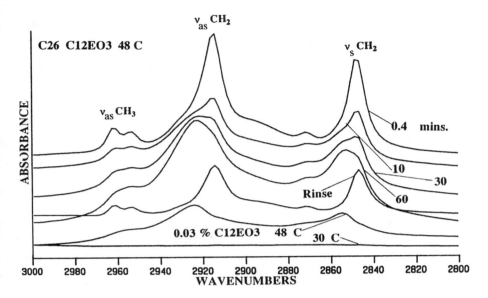

Figure 26. Time - resolved spectra from $C_{26}/C_{12}EO_3$ run at 48 °C at times indicated; After rinsing 15 minutes with water at 48 °C; reference spectra of 0.03% $C_{12}EO_3$ at 48 °C; at 30 °C. All to same scale.

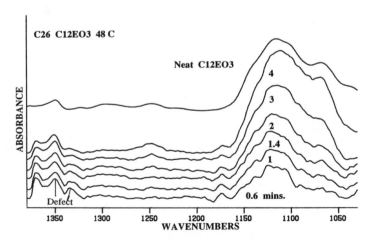

Figure 27. Time - resolved spectra from $C_{26}/C_{12}EO_3$ run at 48 °C. Top spectrum is neat $C_{12}EO_3$ reference on IRE. All to same scale except $C_{12}EO_3$ ref.

by the partitioning of the surfactant between the oil and water phases. The removal mechanisms involved in liquid soil removal are quite distinct in terms of kinetics and transport phenomena from that of the solid soils investigated herein.

Acknowledgments

The author wishes to thank Mr. Daniel Webster for his skillful operation of the spectrometer.

Literature Cited

1. Schwartz, A. M. In *Surface and Colloid Science;* Matijevic, E., Ed.; Wiley : New York, 1972; Vol. 5
2. Schick, M. J. In *Surfactant Science Series*; Marcel Dekker: New York, 1987; Vol. 23
3. Raney, K.H.; Miller, C.A.; Benton, W. J.; *J. Coll. Int. Sci.* **1987**, 117, 282
4. Cox, M. F., *J. Am. Oil Chem. Soc.* **1986**, 63, 559
5. Scheuing, D. R.; Hsieh, J.C.L., *Langmuir* **1988**, 4, 1277
6. Raney, K. H.; Miller, C. A.; *J. Coll. Int. Sci.* **1987**, 119, 539
7. Kahlweit, M.; Strey, R.; Firman, P.;Haase, D.; Jen, J.; Schomacker, R.; *Langmuir* **1988**, 4, 499
8. Kahlweit, M.; Strey, R.; Scomaker, R.; Haase, D.; *Langmuir* **1989**, 5, 305
9. Miller, C. A.; *Colloids and Surfaces* **1988**, 29, 89
10. Schambil, F.; Schwuger, M. J.; *Coll. Poly. Sci.* **1987**, 265, 1009
11. Lindman, B.; Backstrom, K.; Engstrom, S.; *Langmuir* **1988**, 4, 872
12. Scheuing, D. R. *Appl. Spec.* **1987**, 41, 1343
13. Harrick, N. J. *Internal Reflection Spectroscopy;* Interscience: New York, 1967
14. Iwamoto, R.; Ohta, K. *Appl. Spec.* **1985**, 39, 418
15. Sperline, R. P.; Muralidharan, S.; Freiser, H. *Langmuir* **1987**, 3, 198
16. Fink, D. J.; Gendreau, R. M. *Anal. Biochem.* **1984**, 139, 140
17. Snyder, R. G. *J. Chem. Phys.* **1967**, 47, 1316
18. Maroncelli, M.; Qi, S. P.; Strauss, H. L.; Snyder, R. G. *J. Am. Chem. Soc.* **1982**, 104, 6237
19. Snyder, R. G.; Schactschneider, J. H. *Spectroschim. Acta* **1963**, 19, 85
20. Snyder, R. G.; Maroncelli, M.; Strauss, H. L.; Elliger, C. A.; Cameron, D. G.; Casal, H. L.; Mantsch, H.H.; *J. Am. Chem. Soc.* **1983**, 105, 134
21. Precht, D.; Frede, E.; *Acta Cryst.* **1983**, B39, 381
22. Scheuing, D. R.; *Langmuir* **1990**, 6, 312
23. Casal, H. L.; Mantsch, H. H.; *Biochim. Biophys. Acta* **1984**, 779, 381
24. MacPhail, R. A.; Snyder, R. G.; Strauss, H. L.; *J. Am. Chem. Soc.* **1980**, 102, 3976
25. Cameron, D. G.; Casal, H. L.; Mantsch, H. H.; *Biochemistry 1980*, 19, 3665
26. Kawai, T.; Umemura, J.; Takenaka, T.; Gotou, M.; Sunamoto, J. *Langmuir* **1988**, 4, 449
27. Cameron, D. G.; Umemera, J.; Wong, P.; Mantsch, H.H. *Colloids and Surfaces* **1982**, 4, 131
28. Mantsch, H. H.; Kartha, V. B.; Cameron, D. G. In *Surfactants in Solution*; Lindman, B. ,Mittal, K. Eds.; Plenum Press: New York, 1984; Vol 1, p 673
29. Kawai, T.; Umemura, J.; Takenaka, T.; Kodama, M.; Seki, S. *J. Coll. Int. Sci.* **1985**, 103, 56
30. Umemura, J.; Kawai, T.; Takenaka, T.; Kodama, M.; Ogawa, Y.; Seki, S. *Mol. Cryst. Liq. Cryst.* **1984**, 112, 293
31. Snyder, R. G. *J. Chem. Phys.* **1979**, 71, 3229

RECEIVED August 2, 1990

Author Index

Affiliation Index

Subject Index

P

Passivation, copper and nickel, 222
Path length, calibration, DIT–NDX
    method, 79
pH effects, solutions of
    alkyldimethylamine oxide
    surfactants, 123–142
Phase behavior, aqueous surfactant
    systems, infrared
    microspectroscopy, 71–85
Phase behavior of lipids
    cholesterol-induced changes, 56–69
    factors determining, 5
Phase diagram(s)
    aqueous surfactant systems, 71–73
    decyldimethylphosphine oxide–water
        system, 74f
    temperature versus pressure, sodium
        oleate system, 54f
Phase inversion temperature
    efficiency of mechanical agitation, 252
    rates of oil solubilization and removal
        from fabrics, 252
Phase structure, 83
Phase studies
    boundary definition, 83
    decyldimethylphosphine oxide, series of
        spectra, 79–81,82f
    infrared microscopy, model system, 72–85
    liquid crystal boundary definition, 83
    liquid phase boundary definition, 83
Phase transitions
    FTIR spectroscopy, 87
    hydrated solid to micelle, thermally
        induced, 87
    micellar surfactant solutions, 87–120
    micelle to coagel, high pressure, 87
    monomer to micelle, concentration
        dependent, 87
Phospholipids
    analysis, 29–31
    FTIR spectroscopy, application, 4–5
    Raman spectroscopy, application, 4–5
    structure, 4–5f
Phospholipid acyl chain conformational
    disorder, quantitative determination
    using FTIR, 24–42
Phospholipid bilayer systems,
    conformational disorder, 28

Phospholipid gel to liquid crystal phase
    transitions, 25–26
Phospholipid monolayer films,
    thermodynamic phase
    transition, 192–207
Phospholipid synthesis
    DO2/lipid ratio, 29
    hydration of the phospholipid, 29
Photometric inaccuracies, optical
    considerations, 73–75
Physical state of the bilayers
    function of cholesterol
        concentration, 65–69
    temperature dependence, 65–69
pK_a, amine oxides, 126
Plane electromagnetic wave in a
    three-phase system, 199f
Plasma proteins, adsorption on
    biomaterials, 223
Plateau region, molecular order, 64
Polar groups, cholesterol effects, 62
Polar headgroup, inductive effect on
    methylene chain, 135
Polymer(s), 2
Polymer adsorption, 15–16
Polymeric Langmuir–Blodgett films, FTIR
    and gas transfer studies, 177–19
Polymerized lipid, molecular
    structure, 178,180f
Polysaccharide(s)
    adsorption at aqueous–solid
        interfaces, 208–224
    adsorption on germanium, 216,218f,219
    retention at aqueous–solid
        interface, 223
Porous substrate, Langmuir–Blodgett
    films, 177,178
Preformed polymerized lipids, multilayer
    structures, 178
Pressure, effect on the Raman intensity
    ratio of the alkyl chains of sodium
    oleate, 53f
Pressure-induced phase transition,
    temperature effects, 51–53
Pressure measurement in optical cells, 46
Pressure tuning of spectra, 44–54
Protein(s)
    adsorption
        adsorption-induced changes in
            secondary structure, 13–14

Production: Peggy D. Smith
Indexing: Julie Poudrier Skinner
Acquisition: A. Maureen Rouhi

Books printed and bound by Maple Press, York, PA

Paper meets minimum requirements of American National Standard
for Information Sciences—Permanence of Paper for Printed Library
Materials, ANSI Z39.48–1984   ∞

Other ACS Books

Chemical Structure Software for Personal Computers
Edited by Daniel E. Meyer, Wendy A. Warr, and Richard A. Love
ACS Professional Reference Book; 107 pp;
clothbound, ISBN 0–8412–1538–3; paperback, ISBN 0–8412–1539–1

Personal Computers for Scientists: A Byte at a Time
By Glenn I. Ouchi
276 pp; clothbound, ISBN 0–8412–1000–4; paperback, ISBN 0–8412–1001–2

Biotechnology and Materials Science: Chemistry for the Future
Edited by Mary L. Good
160 pp; clothbound, ISBN 0–8412–1472–7; paperback, ISBN 0–8412–1473–5

Polymeric Materials: Chemistry for the Future
By Joseph Alper and Gordon L. Nelson
110 pp; clothbound, ISBN 0–8412–1622–3; paperback, ISBN 0–8412–1613–4

The Language of Biotechnology: A Dictionary of Terms
By John M. Walker and Michael Cox
ACS Professional Reference Book; 256 pp;
clothbound, ISBN 0–8412–1489–1; paperback, ISBN 0–8412–1490–5

Cancer: The Outlaw Cell, Second Edition
Edited by Richard E. LaFond
274 pp; clothbound, ISBN 0–8412–1419–0; paperback, ISBN 0–8412–1420–4

Practical Statistics for the Physical Sciences
By Larry L. Havlicek
ACS Professional Reference Book; 198 pp; clothbound; ISBN 0–8412–1453–0

The Basics of Technical Communicating
By B. Edward Cain
ACS Professional Reference Book; 198 pp;
clothbound, ISBN 0–8412–1451–4; paperback, ISBN 0–8412–1452–2

The ACS Style Guide: A Manual for Authors and Editors
Edited by Janet S. Dodd
264 pp; clothbound, ISBN 0–8412–0917–0; paperback, ISBN 0–8412–0943–X

Chemistry and Crime: From Sherlock Holmes to Today's Courtroom
Edited by Samuel M. Gerber
135 pp; clothbound, ISBN 0–8412–0784–4; paperback, ISBN 0–8412–0785–2

For further information and a free catalog of ACS books, contact:
American Chemical Society
Distribution Office, Department 225
1155 16th Street, NW, Washington, DC 20036
Telephone 800–227–5558